CAMBRIDGE LIBRARY COLLECTION

Books of enduring scholarly value

Life Sciences

Until the nineteenth century, the various subjects now known as the life sciences were regarded either as arcane studies which had little impact on ordinary daily life, or as a genteel hobby for the leisured classes. The increasing academic rigour and systematisation brought to the study of botany, zoology and other disciplines, and their adoption in university curricula, are reflected in the books reissued in this series.

Where There's a Will, There's a Way!

During the eighteenth and nineteenth centuries many scientists, naturalists, engineers and inventors from humble backgrounds, largely self-taught, made significant contributions to British science. This 1873 book by James Cash (1839–1909) celebrates their achievements in natural history, while promoting a 'self-help' ideology, stressing how disadvantages could be overcome by those with ability and determination. Many of his subjects corresponded with great names such as William Jackson Hooker, and sent specimens or local information which helped build up the larger picture. Cash gives particular attention to men from the north of England, where many men engaged in the cloth trade were also notable plant collectors. His subjects include George Caley, a weaver self-taught in Latin and French, and whom Sir Joseph Banks employed to go to New South Wales as a collector; Edward Hobson, a factory worker; and John Horsefield, a self-taught weaver who memorised the Linnaean orders at his loom.

Where There's a Will, There's a Way!

Or, Science in the Cottage; An Account of the Labours of Naturalists in Humble Life

JAMES CASH

CAMBRIDGE UNIVERSITY PRESS

Cambridge, New York, Melbourne, Madrid, Cape Town,
Singapore, São Paolo, Delhi, Tokyo, Mexico City

Published in the United States of America by Cambridge University Press, New York

www.cambridge.org
Information on this title: www.cambridge.org/9781108037907

This edition first published 1873
This digitally printed version 2011

ISBN 978-1-108-03790-7 Paperback

WHERE THERE'S A WILL THERE'S A WAY!

OR,

SCIENCE IN THE COTTAGE:

AN ACCOUNT OF THE

LABOURS OF NATURALISTS IN HUMBLE LIFE.

By JAMES CASH.

LONDON:
ROBERT HARDWICKE, 192, PICCADILLY.
1873.

LONDON:
PRINTED BY WILLIAM CLOWES AND SONS,
STAMFORD STREET AND CHARING CROSS.

PREFACE.

The object of this work is, briefly, to record the labours, and bear some testimony to the character, of a class of men, for the most part in the humblest walks of life, who have rendered no mean service to science, and whose memory it is desirable to rescue from oblivion. Of men gifted with splendid genius, whose works are known the world over, and whose names will ever live as those of scientific investigators and discoverers, it is not the Author's purpose to speak. His attention has been confined (except as to two individuals, whom he found it impossible to overlook) strictly to men in humble life—"working men," as the phrase is—who, earning their livelihood, and that, in most cases, an extremely precarious one, by hard daily toil, yet found leisure for the practical study of Natural History. In spite of social and educational disadvantages—in spite even of deep poverty—they pursued their labours with indomitable perseverance and devotion. How true it is, that—

> " Strongest minds
> Are those of whom the noisy world
> Hears least!"

The position of men working at the loom, in the mechanic's shop, or at the shoemaker's bench for ten or twelve hours a day, is not favourable to study of any kind; yet, in proportion to the magnitude of their difficulties, the greater was the ardour, the more unyielding the determination, of these self-taught naturalists. Their enthusiasm maintained even to old age—demanding, in some cases, enormous sacrifices—carried them to the utmost limit of physical endurance. Fatigue was willingly, nay joyfully, endured, for the sake of the pleasure which accompanied it, or followed in its train. To many the capture of a rare insect, or the finding of a new plant, was often the best and only reward of a toilsome walk of thirty miles. They followed science most truly for its own sake.

It is gratifying to know that the labours of the humble Botanists who flourished, chiefly in Lancashire, during the first half of this century, received some recognition from the eminent men who ranked as leaders and discoverers in Natural Science. Dr.

Withering, Sir James Edward Smith, Sir William Jackson Hooker, Dr. Greville, and many others, knew them and learned to esteem them. No disparity of social rank was a hindrance to free communication with them. The importance of a due recognition of genius in the humbler walks of life was often urged by the late Professor Sedgwick. "I wish," he once said, "that the barriers between man and man—between rank and rank—should not be harsh, and high, and thorny; but rather that they should be a kind of sunk fence, sufficient to draw lines of demarcation between one and another, and yet such that the smile of gladness and the voice of cheerfulness might pass over and be felt and heard on the other side." No one knew better than the Professor how to sink social differences in the pursuit of higher knowledge.

The honest labours of these "Naturalists in Humble Life," deserve to be placed on record, not only in the interest of science, but also as an example to many—it is hoped very many—whose tastes may lead them into similar paths.

The Author, in addition to extended and careful personal investigation has availed himself of such fragments of information as have appeared in print from time to time; and he sends this result of his labours forth as a story of plain men, plainly told. Whatever be the faults of compilation, he would beg the reader not to think lightly of the struggles after self-culture of weavers, factory-hands, and shoemakers, but, making some allowance for the circumstances of individuals, award them a measure of praise for that which they have achieved.

The old Lancashire Naturalists were respected, not more for their attainments, than for their personal excellence. They were, on the testimony of gentlemen high in position, as pure and blameless in their lives as, when advanced in years, they were disposed to bear, without repining, the changes and privations to which age and infirmity subjected them.

The only naturalist now living whose career is noticed at length, in these pages, is Thomas Edward, A.L.S., of Banff. To him the Author would direct especial attention as one of the most remarkable scientific men of modern times.

Longsight, Manchester,
November, 1873.

CONTENTS.

WHERE THERE'S A WILL THERE'S A WAY.

EARLY ENGLISH BOTANISTS.

THE study of botany by working men, chiefly in the North of England, commenced in the pre-Linnæan era, of which the records are extremely scanty. That study may not, however, have been conducted in the most scientific manner. There is something in plants, apart from scientific investigation, something in the associations which cling to them, and in the localities they inhabit—something, too, in the memories they recall—which gives them a place in the regard of everyone; but the aids to scientific study, before the sexual system of Linnæus made botany easier of comprehension to ordinary minds, must have been slight indeed. If plants in that age were endowed with impossible virtues, if popular belief bestowed on them extraordinary powers of development by unheard of processes, some excuse may be urged, for it was an age when even the wealthier classes were little better educated than many of the working class now are.

The names of working men, in connexion with systematic botany, first appear in the time of Ray—

that is, about the middle of the seventeenth century. Associated with that great botanist, and sharing in a humble way his reputation, was Thomas Willisel, "an unlettered man"—to use the words of Dr. Pulteney—but one to whose love for plants, and his zeal and assiduity in collecting them, testimony is borne by the writers of his day. It is on record that Ray and Willisel "travelled through Derbyshire, Yorkshire, and all the Northern Counties as far as Berwick; and back through the Bishopric of Durham," investigating the botany of the country. Much of the personal history of Willisel, all, in fact, that does not bear directly on his connexion with Ray, and his contemporaries Merrett and Morison, is wrapt in obscurity. The historian of early English botany, already quoted (Dr. Pulteney), expresses regret at being unable to commemorate the name of Thomas Willisel, since he is "uninformed of the time and place, both of his birth and of his death." He was, however, a remarkable instance of what a man may achieve by his own indomitable energy. His whole life almost was devoted to the study and collection of plants; and, living as he did in an age when the botany of his native country had been but little studied, he was able to add largely by his travels to the stock of new discoveries. Not only Ray, but Dr. Morison also, the author of the 'Historia Plantarum Oxoniensis'—soon after his settlement at Oxford—employed Willisel to collect rare English plants; and for five summers he travelled at the expense of Dr. Merrett into different parts of England with the same object, and especially to make collections for the 'Pinax,' which is said to have been enriched by many of the rarer species by the labours of Willisel. There is a presumption, also, that this man was sent into Ireland by Dr. Sherard.

It may be objected that Willisel pursued his botanical labours from a mercenary motive. If so, let the objector consider for a moment the state in which botanical science was in his day. Not only were a large proportion of the plants growing in Great Britain then unrecorded, but the science itself, so far as classification and arrangement were concerned, was in a nascent state. Ray and his learned contemporaries were endeavouring to reduce to order that which they had found in a state of chaos. It was impossible that, by their own unaided efforts, they could acquire a full knowledge of the indigenous plants of this country; and, when a man like Willisel—whose genius cannot be doubted—offered, or was willing to travel through the country at their expense, it is the most natural thing in the world that they should have employed him for the purpose of collection.

It is not to be supposed, moreover, that Ray and Morison were idle, or that they confined their investigations to the botany of the neighbourhoods of Cambridge and Oxford. Ray, after publishing his Catalogue of plants around Cambridge, 'Catalogus

Plantarum circa Cantabrigiam,' in 1660, which included 626 species—resolved to prepare materials for a more pretentious work. He had already in 1658 made a tour alone of the Midland Counties; and in 1861 he spent six weeks, from July 26th to August 30th, upon a similar tour through the Northern Counties. In the following year, accompanied by Willoughby, he took a third and more extensive tour through the Midland Counties into Cheshire; thence into North Wales and the West of England. This excursion occupied from May 8th to July 18th, and Ray gathered "a plentiful harvest," which afterwards enabled him to enrich his general 'Catalogue of English Plants,' then in meditation. In the autumn of 1668, he took a fifth journey alone into Yorkshire and Westmorland. It is certain that in the course of these and other journeys—especially in the Northern Counties—Ray met with many men noted for their local botanical knowledge. We find him, in the preface to his 'Synopsis,' which appeared in 1690, acknowledging the services amongst others of "Thomas Lawson, of Strickland, in Westmorland," and James Newton, "a diligent and skilful botanist," both, no doubt, men in humble life.

A great stride in the knowledge of British plants was made in the lifetime of Ray, and chiefly through his labours. "In the space of little more than twenty years," says Dr. Pulteney, "he had seen the English flora acquire an accession of upwards of

500 new species." His 'Catalogus Plantarum Angliæ' (1670) contained about 1050 species; and in the second edition of his 'Synopsis,' which was given to the world in 1696, he enumerated fully 16,000 species, for the discovery of many of which credit is certainly due to Thomas Willisel.

Willisel's knowledge was not confined to the vegetable kingdom. His patron informs us that "he was employed by the Royal Society in the search of natural rarities, both animals, plants, and minerals, for which purpose he was the fittest man in England, both for his skill and industry." In Ray's correspondence there occurs an observation made by Willisel of the various trees on which the mistletoe was found growing.

From the death of Ray and his contemporaries, until the time of Dillenius, botanical science—except so far, in this country, as it was fostered by Sir Hans Sloane and others—seems to have languished. Dillenius, during his residence in England, was no less diligent in searching for rare plants than were his eminent predecessors; and he adopted a similar plan, namely, that of employing men in a comparatively humble position to collect. One of these was Samuel Brewer, whose researches were chiefly among the cryptogamic tribes, and who, during the preparation of the 'Historia Muscorum,' sent "great numbers" of scarce plants to Dillenius. Brewer devoted himself to science, after an unsuccessful attempt to establish a woollen manufactory at Trowbridge—of which place he was a native—and he was one of Dillenius's earliest correspondents. In the special department which he undertook, he accomplished much by his skill and assiduity. The pursuit of English botany amounted to a passion with him. In the summer of 1726, he attended Dillenius in a tour through Wales, Anglesea, and the Isle of Man. For the purpose of collecting mosses, of which the larger portion attain their development in winter, Brewer spent the winter of 1726–27 in Wales, and made frequent excursions among the mountains, chiefly about Snowdon. Having received some instruction from Dillenius, he was enabled to detect many new or rare species; and such plants as he had no previous acquaintance with he sent to Dillenius to determine the species. "I have seen," says Dr. Pulteney, writing more than half a century afterwards, "a catalogue of more than 200 plants—many of which were ill-ascertained before—all sent at one time, with references to the 'Synopsis' affixed by Dillenius."

Brewer settled at Bradford, Yorkshire, and spent there the later years of his life. He collected materials for publishing what was to have been called a 'Botanical Guide,' but it never saw the light. He died about the year 1742.

Among the amateur botanists of the period referred to was Thomas Harrison, of Manchester. He was a tradesman; but mention of his name ought not on that account to be omitted. Like Brewer, he furnished Dillenius with specimens. A herbarium of British and exotic plants which he collected is said to have contained in 1762 no fewer than 4000 specimens. It was richest in exotic *Filices*. This herbarium, after the collector's death, was purchased for a considerable sum and deposited in the Manchester Library.

John Wilson, a native of Kendal, is quoted by the historian of English botany as being, like Willisel and Brewer, "an instance of that unconquerable attachment to a favourite branch of knowledge which sometimes engrosses the minds of those who, by their lot, have not been exempted from labouring in the lower and mechanical offices of life." This man, who was, according to one account, a stocking-maker, must have acquired an intimate knowledge of botany, for he is said to have given weekly lessons in that science alternately in his native town and at Newcastle-upon-Tyne. He made about £60 a-year by teaching — no inconsiderable income in those days—and so high was his reputation that many pupils resorted to him from the south of Scotland. Other accounts, however, represent Wilson as having followed the trade of a shoemaker, and lived during a great part of his life in a state of indigence. A story is related of him as showing his enthusiasm for botany, which is hardly credible. It is this: that being in want of Morison's 'Historia Plantarum,' and having no other means of procuring it, he determined to dispose of almost the only means of subsistence which his family had—a cow—and devote a large part of the proceeds to the purchase of the coveted volume. A benevolent lady in his neighbourhood, hearing of this, presented him with the book, and thus rescued the poor man and his family from ruin.

In 1744 Wilson published 'A Synopsis of British Plants in Mr. Ray's method; with their characters, descriptions, places of growth, time of flowering, and physical virtues, according to the most accurate observations and the best modern authors; together with a Botanical Dictionary, illustrated with several figures.' This was an octavo volume of 272 pages, and was published at Newcastle-upon-Tyne. The author died six years after the appearance of his book. What was to have formed a supplement to it, or a second volume, treating of the Gramineæ and Cryptogamic plants of Britain, he left in manuscript.

Distinguished as were the labours of some of the pre-Linnæan botanists, they nevertheless, owing to the want of a recognised system of classification, groped to a large extent in the dark. So, at least, would it appear to botanists of the nineteenth century, to whom, such has been the advance in knowledge, the Linnæan or sexual system itself is antiquated. The introduction of this system, which was shadowed forth in the 'Florula Laponica' (1732), and in the first sketch of the 'Systema Naturæ,' and afterwards more clearly exemplified in the 'Genera

Plantarum' of Linnæus, was as the dawning of a brighter day; and when, in the course of years, it came to be recognised as a comprehensive and tangible thing, men to whom, on account of their limited means, botany had been a sealed book, found it to be the key by which they were enabled to unlock the treasury of Nature, and find stores of genuine gratification.

Moved by his esteem for the character of Dillenius, then residing in England, and also by a desire to inspect the 'Pinax' of Sherard, which is said to have been not the least among his motives, the Professor of Upsal visited this country in 1736. The elder botanists, who were averse to change, received him coldly. The system, however, which will ever be associated with his name, was destined ere long to supplant those of Ray and Morison; and the universal acceptance which it met with, ere he died, was its author's best reward. Henceforward an impulse was given to botanical studies which had no parallel in the previous history of the science. Its influence is felt still, and will continue to be felt, probably, for all time.

BOTANY IN LANCASHIRE.

NOTHING could be more remarkable than the way in which a love for plants was developed among the operatives of Lancashire towards the close of the last, and during the first half of the present, century. There is reason to believe that, in the time of Dillenius, when Brewer was exploring the mountains of Wales, or, later, when John Wilson was reaping a rich botanical harvest in Westmorland, botanical science was methodically studied by men of humble position in Lancashire. It is not, however, until about the year 1770, when the fame of Linnæus had become established, and his system had received general acceptance, that work was done of which we have any record. The accounts of the amateur botanists of that era are extremely meagre. It is only when they began to band themselves together and form local societies—an evidence of the popularity of the pastime—that we hear anything of their movements; for, except in the case of George Caley, to be hereafter mentioned, none, of whom any record remains, was, like Willisel, largely employed as a collector. Moreover, the self-imposed task of these men—the task of ascertaining, with a minimum of opportunity and means, the botanical features of their respective neighbourhoods—was carried on in such an unassuming manner that, however interesting their discoveries, or however valuable for comparison, a knowledge of the Manchester flora of a hundred years ago might now have been, no memoranda made at that time are known to exist.

The village of Eccles, six miles from Manchester, was, as far as is known, the first botanical centre in Lancashire.

James Crowther, one of the most noted of the Manchester botanists, is said to have attended, about the year 1777, the meetings of a society there which numbered about forty members. A society so numerically strong, holding its meetings in a small agricultural village, could not have been the creation of a day; and it is not improbable, although there is no record of the fact, that it had its origin shortly after the visit of Linnæus to this country. No doubt the publication of the 'Systema Naturæ' and 'Genera Plantarum'—works which the early Lancashire botanists deemed of great value—gave an impulse to their studies, and led them into association. At Oldham, Ashton-under-Lyne, and other towns too remote from Eccles to admit of the meetings there being conveniently attended, for every journey had to be performed on foot, district botanical societies were established; and, in order that the members might meet for mutual encouragement, instruction, and a comparison of discoveries, general conversazioni, to which all were invited, were held at intervals.

A far-reaching organisation like this must have taken years to develope. The collective meetings were held, according to a pre-arranged plan, in some respectable public house, and were long conducted without any approach to licence or disorder; but, about the close of the century, "on account," as we are informed, "of infractions of the rules," they were discontinued. Still the district societies grew and multiplied. There was no abatement of energy on the part of individual botanists. Much valuable work was done; and those humble votaries of science were not without a reward in the gratification that work afforded them. As a rule their regular meetings were held—as those of some societies still are—on Sunday. This will appear, to many, a highly objectionable circumstance; but, in fairness, it should be remembered, that Sunday was the only day when these hard-working sons of toil could by any possibility meet for the interchange of ideas. Let us hope that they met in the spirit that is breathed through 'The Artizan's Song,' by John Critchley Prince—one of the band of *poets* in humble life who are the pride of Lancashire—a man who, after charming the fireside of many an artizan by his beautiful lyrics, died, only a few years ago, in the lowest depths of poverty:—

"I am glad when the Sabbath steals quietly in,
Of all days the chief lustre, the pearl of the seven;
A season when man seems to pause in his sin,
A time, rightly used, giving glimpses of Heaven!

"Then I seek, with my household, the Temple of man,
And to God offer up my own heart-utter'd prayer;
But believe me not lost if I go, now and then,
To the Temple of Nature, and worship Him there."

The general botanical meetings were resumed about the year 1810, under the presidency of John Dewhurst; and in more recent years they have had their counterpart in the annual gatherings of the Amalgamated Botanical Societies of Lancashire, Cheshire, and Derbyshire.

At the earliest period of their association, the Manchester botanists, out of their modest earnings, subscribed to form a library of reference, which, even in Crowther's boyhood (he was born in 1768), must have been of considerable value. Amongst other works which it contained on the discontinuance of the collective meetings, about seventy-five years ago, were the 'Species Plantarum' and 'Systema Naturæ' of Linnæus, and several works by English authors, which, though of relatively smaller value, had done good service. Dr. Withering was a favourite author. His 'Arrangement of British Plants,' and other works, as well as the works of Dr. Hull—a Manchester physician, who published a British Flora in 1799—and the 'Flora Anglica' of Hudson, were for a long time kept in circulation, and systematically studied.

The above facts show how sound and honest the labours of these men were, and that they did not look at plants merely from an utilitarian, or herbalist's, point of view.

Testimony is thus borne to their character in the 'Magazine of Natural History' (1829):—"No person who is not conversant with some one belonging to the Manchester school of botanists can have any conception of the ardour and devotedness with which the discovery and cultivation of the rarer species is pursued. They live in a little world of their own, are remarkable for their sobriety and industry, and derive many advantages from that sectarian spirit which, separating the few upon some distinct ground from the many, obliges them to uphold each other, and magnify the importance of the object which associates them."

Similar testimony was borne to the character of these men by Professor Sedgwick, on the occasion of the visit to Manchester of the British Association for the Promotion of Science, in 1842. At the Association dinner, the learned professor proposed, in the happy strain which characterised all his public utterances, "The Literary and Philosophical and other scientific societies of Manchester." He spoke of a walk that he had had through the streets of Manchester, and said:—"In going through those dark corners, in talking to men whose brows were smeared with dirt and whose hands were black with soot, I found upon them, nevertheless, the marks of intellectual minds and the proof of high character; and I conversed with men who, in their own way and in many ways bearing upon the purposes of life, were far my superiors. I would wish the members of the British

Association to mingle themselves with these artizans in these, perhaps, unlooked-for corners of our great cities; for, as I talked with them, the feeling prevailing in my mind was that of the intellectual capacity manifested in the humbler orders of the population in Manchester. Whilst the institutions and customs of men set up a barrier, and draw a great and harsh line between man and man, the hand of the Almighty stamps His first impress upon the soul of many a man who never rises beyond the ranks of comparative poverty and obscurity. Hence arises a lesson of great practical importance—that we should learn in our walks through life, in our mingling with the busy scenes of the world, a lesson of practical wisdom, of kindness, of humility, and of regard to our fellow beings."

JOHN DEWHURST: POLITICAL TROUBLES.

JOHN DEWHURST, of Manchester, who followed the precarious trade of a fustian cutter, has been mentioned as the first president of the collective meetings of the Lancashire Botanists, held after the revival of the central organisation, in the early part of the present century. Of this man hardly anything is known; but that he was an excellent botanist is certain from the position he held, and from the casual references to him which appear in the letters or memoranda of those who knew him personally. Though his senior by many years, he was the frequent companion of Edward Hobson, the muscologist, than whom none of the Manchester botanists enjoyed a higher reputation, and the two were associated in the management of a small private botanical garden, given to them by Mr. Mitchell, of Bradford, near Manchester. In this ground the less common plants which they gathered in course of their rambles were cultivated.

In order to comprehend the difficulties under which Dewhurst and his associates laboured, it is only necessary to recall the history of that period. It was a period of universal disquiet, and, in this country, of widespread distress.

"Many rich
Sank down, as in a dream, among the poor;
And of the poor did many cease to be,
And their places knew them not."

The French Revolution had been accomplished; Napoleon was pursuing his career of conquest on the Continent; and, a military fever having seized on the people of England, English armies were sent to cope with him. By none more than the working classes, in their ignorance of its consequence, had that war fever been encouraged. They early suffered the effects of their folly in paralysed trade, exorbitant taxation, and, as a matter of necessity, semi-starvation. The unhappy weavers of Lancashire felt these effects most dreadfully, and, chafing under a state of things which they were powerless to remove, vented their feelings often in riot and sedition. It is not too much to say that the experience of the first twelve years of this century converted Lancashire, which had been a stronghold of Toryism, and Manchester, which had been the most Conservative town in the kingdom, into the most Radical; for it was there that the reform agitation, timidly commenced, was afterwards most vigorously carried on. It was there, also, and in the immediate vicinity, that the most serious disturbances occurred in consequence of the depressed condition of the staple trade, which was itself a natural result of the prolonged war with France. It would be foreign to the purpose of this work to mention in further detail the circumstances of that eventful period. The reader will be desirous of knowing whether, and to what extent, the botanists of the district were concerned in the politics of the time.

Many of the early botanists of Lancashire were weavers, or were engaged in some other branch of the cotton industry, or in some humble occupation which could not escape the general depression; yet the records of that period will be searched in vain for the name of any naturalist (as far as their names are known) among the hundreds of men who were imprisoned for riot or conspiracy. Their habits were essentially peaceful. Whilst the operatives of Middleton were sacking and burning the cotton mills which had ceased to yield them employment, and whilst soldiers were shooting down the rioters without mercy, Dewhurst, Hobson, Horsefield, Mellor, Crowther, and a host of others—all worthy men—were pursuing their harmless and ennobling pleasures upon the Lancashire hillsides, or in the fields, meadows, and lanes of the neighbouring county of Cheshire. Dewhurst and Hobson only, of the entire band, seem to have come even under suspicion—whether deservedly or not it is impossible to say. Those were days when for men to meet even for a lawful purpose—to discuss actual grievances, and to demand a remedy for acknowledged abuses—was to expose them to suspicion and almost certain imprisonment. It is recorded that one day in 1812 these men—Dewhurst and Hobson—then inseparable companions, had engaged to attend a weavers' meeting, which the local justices had made up their minds to disperse. Mr. Mitchell, the gentleman to whom the two botanists were indebted for their slice of ground, had private intimation of the justices' intention to arrest the leaders, and as it happened to be the day when Dewhurst and Hobson visited their garden, as usual, he pressed them to remain, and, it is said, even plied them with drink in order to keep them out of the way of worse harm. Hobson—on what grounds is not known—was one of the marked, and had he fulfilled his intention and gone to the meeting, would certainly have been arrested. On his way homeward with Dewhurst, he met the soldiers with a number of men in custody on their way to the gaol. No doubt he was happy at the thought of his escape, and took warning.

John Horsefield took an interest in politics, though in a quiet way. He wrote: "With politics I had little to do practically; but in 1816, after the war was over—an event which I had been taught to look to for the restoration of good trade—finding ourselves disappointed, that year of dear provisions and no work turned all my neighbours, as well as myself, into a kind of politicians. 'Parliamentary Reform' was the cry. I attended several reform meetings in 1816, and for a few years after. I attended one at Bury and one at Middleton, at each of which Samuel Bamford presided. I was at Peterloo, Manchester, on the 16th August, 1819,* as a spectator; but had nothing to do with any procession. I was there at noon; and Mr. E. Hobson, who then lived in Tassell Street, Manchester, coming on the ground in the dinner hour saw me. 'Well, John,' he said, 'are you here?' — 'Yes.' 'Come; go with me and get your dinner.' — 'No,' I said, 'I think of stopping here, and watching the proceedings.' But I afterwards regretted that I did not go with him, when I had to leave the ground so precipitately. That was the last political meeting I ever attended."

These incidents serve to show the character of the times in which very many of the naturalists, of whom we have to give an account, lived.

* A great reform meeting, held on St. Peter's Field (now the site of the Free Trade Hall), Manchester, was dispersed by the military, and, from the slaughter which took place, the event acquired the name of "Peterloo." "Eleven persons were killed, six hundred wounded, and sixty thousand carried to their homes the recollection of that fatal day."—*Prentice's History of Manchester.*

GEORGE CALEY.

ONE of the most zealous of the botanists who flourished in Lancashire about the beginning of this century was George Caley, the son of a Yorkshire horse dealer, who, when George was a boy, removed to Middleton, near Manchester, and engaged in farming. Caley's career was a remarkable one, for not only did he associate with the local botanists, among whom he held high rank, but he also corresponded and maintained a close acquaintance with Dr. Withering, Dr. Hull, and other botanical celebrities; and, moreover, finding his native country too limited for the exercise of his genius, he traversed a considerable part of the globe in quest of plants.

The circumstances of the early life of George Caley were not very propitious. After his father's settlement at Middleton, he was sent to, and spent a period of four years at, the Manchester Free Grammar School; but at the age of twelve, when about to enter upon a classical course, his father deeming the acquisition of Latin unsuited, or at least unnecessary, to the future occupation, of the boy, took him from school and put him to the stable. He was not long, however, suffered to remain there. He had not obtained such a knowledge of arithmetic as his father deemed necessary; and he was, therefore, sent back to school. In a few months—perhaps weeks—he was permanently consigned to the drudgery of the farm.

A love for plants, and a desire to familiarise himself with them, seems to have early dawned upon George Caley. His father's design was to train him to his own line of business. The quackery and ignorance of the local farriers did not escape his notice, and finding an odd volume of a work on Farriery upon his father's shelves, containing anatomical figures and recipes for certain diseases, with references to useful plants, he applied himself to the study of the subject. He began to search for the plants named, and had no sooner done so (to use his own words) than he "wanted to know more about them." Step by step, having at his command the slenderest means, and no better botanical guides than some "petty herbals," which failed to satisfy him, he became acquainted with such plants as grew in his immediate neighbourhood. His was literally in those early days a groping in the dark. Linnæus and his system were unknown to him; and Johnson's edition of 'Gerarde,' upon which he stumbled, tended only to perplex him. He had, however, in course of a few years, obtained such a knowledge of plants—perhaps not a systematic knowledge, but at all events one which was the foundation of better things—that nothing could quench his determination to make botany a life's study.

"In this state of things," he says, writing to Dr

Withering (June 15th, 1798), "I heard of your 'Botanical Arrangement.' It was not long before I got a copy of the second edition. I was now at a greater loss than ever; for I really could not tell what to make of it. However, I concluded that it was nonsense to let books lie idle. Winter was approaching, and no plants were to be seen. I resolved to learn the introduction, and soon gained a tolerable idea of it. I then wished to see some flowers, but still the dreary winter was before me. I was obliged to put up with the inconvenience, so I learned the introduction over and over before the spring. When the plants began to flower, I began to try my strength in the science; but, knowing a good many plants before, I used to cover the names, in order that I might not favour one character more than another. Sometimes I was right; sometimes wrong; but by this method I gained a good knowledge of investigation."

The severe winter study of which he here speaks placed Caley on a par, in respect of technical knowledge, with many of his contemporaries who in the enjoyment of ampler means had received definite scientific instruction.

The characteristic ardour of the youth and his determination not to let even the occupation which his father had assigned him interfere with his botanical pursuits is apparent from an incident which he relates in the letter above quoted. "I began," he says, "to find out botanical companions—for, before

I had laboured myself—but they followed some manufacturing employment; and mine would not permit me to spend the time which they did. Hereupon I determined that I would learn to weave." This determination he carried out, but no sooner had he gained a fair idea of handloom weaving than, in consequence of the French war, there occurred a general stagnation of trade: he therefore returned to the stable. In the course of his early botanical investigations, the occurrence in gardens of many exotic plants, of which the English manuals gave no account, puzzled him; and in order to become acquainted with these, he purchased Linnæus's 'Genera Plantarum' and 'Systema Vegetabilium.' The author's Latin terms were a further source of perplexity, for Caley had forgotten most of his school Latin; but as the terms were "technical, and chiefly nouns," he soon recovered that loss.

It was at this time that, in association with the kindred spirits of his neighbourhood, Caley became most assiduous in the cultivation of English botany. Every hill and plain within a day's walk—and that none of the shortest—of Manchester was examined by these men with microscopical diligence; "and Caley," says a writer in the 'Magazine of Natural History,' "was not the person to sink the temperature of his zeal in a subject which wholly engrossed him; and if he received from them some increase of ardour in the first instance, it cannot be doubted but that he transmitted it to them in return with redoubled intensity."

In course of time, there being, apparently, no fresh discoveries to be made at home, Caley entertained an idea of going abroad. How to do this, without ample funds, was a serious problem. First, he thought of becoming a seaman; but then he had had no training, no nautical education whatever —probably had never been on board a ship—and so he abandoned that notion. The step which he did take—a bold one indeed for a Lancashire weaver—was to write to the president of the Royal Society, Sir Joseph Banks, and state his desire to be employed as a botanical collector abroad. He waited long for a reply. At length a reply came. It was, though not in the sense which Caley had hoped, favourable. It did not contain an offer of immediate employment in the direction that he wished; for Sir Joseph, whilst keeping in view the design of his correspondent, did no more than offer him a situation as working gardener in the Botanic Garden at Kew. He knew how valuable would be the experience to be gained in that capacity by one who was entrusted with work abroad. It may be said indeed to have been essential, for at Kew Caley would become acquainted with races of plants, and their habits of growth, of which he must necessarily have been ignorant; and he would acquire in the plant-houses of the Royal Gardens experience of the utmost value. The offer, moreover, was accompanied by a promise of "further help"—a purposely vague phrase—in the event of Caley making "proper progress." He was, to say the least, disappointed, and not a little perplexed as to the course which he should adopt. "I did not," he wrote, shortly afterwards, to Dr. Withering, "much like the thoughts of working in a garden, for that would be out of my element; and being tied to regular hours was not like working piecework, knowing what I had to do, and then giving over." He, however, went to Kew, and was initiated into the practice of gardening. His engagement came to an abrupt termination. "Kew," said he, "was a place which, I sincerely acknowledge, I could not 'weather,' not through the hardship of work, but from being debarred from cultivating my mind according to its natural inclination."

"Alas!" wrote the younger Withering, "the ardent imagination of our aspirant had well nigh outstripped all reasonable expectation; and a series of remonstrances with the president of the Royal Society somewhat injudiciously, not to say intemperately, urged, for a season blasted his high-flown hopes. He whose delight was 'to wander as free as the wind on the mountains,' could ill brook, even for a limited period, the confinement of stated hours, or the restraint of garden walls. After having vainly endeavoured to convince Sir Joseph Banks that he needed no such initiatory course, and that he was already qualified for the projected expedition [to the South Seas], he withdrew in disgust, again to ruminate on his wayward fate amid the wilds of Lancashire." And the writer draws this moral from the conduct of Caley at that time: "Now that talent of every kind is likely to be forced and fostered"—that was in 1830—"perhaps as some may apprehend to an extreme degree, let those whose warm temperaments glow with a laudable desire to excel, beware of yielding to that seductive self-sufficiency which is but too apt to resist the wiser counsels of experience, and thus, in innumerable instances, to make wreck of the brightest expectations. And such disappointments would probably have terminated the career of Caley's usefulness, but that he was so fortunate as to have engaged the attention of patrons not less habituated to detect merit, even through a rough exterior, than to exercise thereon a characteristic generosity and benevolence."

Caley was, at the time of which we write, in correspondence with the elder Withering—the author of the 'Arrangement of British Plants'—who appears to have sympathised with his disappointment. In his letter of the 15th June, 1798, Caley mentions some of the qualifications which he possessed, and the points in his opinion requiring attention on the part of one holding the position he sought. "Having a little idea of manufacturing goods is of great utility to me, for it will cause me to pay attention to plants that are worthy of being applied to such purposes. Those persons who have been sent into foreign parts to collect plants have not favoured agriculture, commerce, and the Materia Medica so much as an inquisitive mind would expect; for they are chiefly such as have worked in some botanic garden." "If," he adds, somewhat bitterly, "I were to mention all the difficulties and fatigues that I laboured under in pursuing my natural inclination, I should never expect to be credited."

Caley was on the friendliest terms with Dr. Withering, and sought his advice in many a difficulty. The physical sufferings of the author of the 'Arrangement' were, his son says, remarkably alleviated by botanical researches; and he was never more agreeably engaged than when fostering rising genius, and especially when promoting the views of the tyro diligently seeking after scientific knowledge, to whom he was ever accessible either by correspondence or personal application. "Among very many who thus benefited by his advice and instructions was George Caley, who, impelled by an ardour sufficient to overcome obstacles and discouragements from which a mind of ordinary temperament would have recoiled, at length resolved to state to him the peculiarity of his situation;" and Dr. Withering "soon became so warmly interested in the welfare of this genuine child of Nature as to continue a correspondence with him during several years, and eventually to assist in advancing his favourite project of exploring the most remote regions of the earth." But the friendship of Dr. Withering did not manifest itself solely in the giving of advice, valuable as that no doubt was. Caley had been recommended to acquire a knowledge, amongst other things, of drawing; and of Latin—both essential to a scientific botanist—and with regard to the former he writes: "I assure you that I have not a good opportunity at present, but on ship-board I intend to try it," showing that then (June 1798) he confidently expected an appointment, if he had not already received one, from Sir Joseph Banks. As to Latin: "I may very easily improve on that, for at the present I understand the declensions of nouns, as well as I did when I went to school, and pretty well of the conjugations of verbs, and also of the agreements of concords, particularly that between the substantive and adjective; or, to speak in short, know what the 'Genera Plantarum' chiefly requires. I think it is not very difficult to learn to read the French, but difficult to pronounce it. Whatever elementary books you would favour me with I could wish to be directed to Strangeways, near Manchester." Such Latin and French elementary books as the younger Withering was possessed of were gladly surrendered for the benefit of the enthusiastic and youthful adventurer; and they were forwarded according to his instructions.

The estrangement between Caley and Sir Joseph Banks was not of long duration, for in July 1798 the former, no doubt regretting the haste with which he had thrown up his post at Kew, wrote to his patron in terms which drew forth the following reply:—

"Soho Square, July 16, 1798.

"Mr. Caley,—Whoever told you that I was angry with you has been mistaken. I am sure I never said so, because I never felt myself angry with you.

"I told you, when I first wrote to you, that unless you would gain your livelihood as a gardener while you made yourself acquainted with plants cultivated in the gardens here, I did not mean to get employment for you as a botanical traveller. By so doing I put you in the same situation as Aiton, Lee, Dickson, and Mason were in when they were of your age; all of whom, at that period, gained their livelihood in the gardens without complaint. No person has been appointed to go to Botany Bay in your stead. The man who is going, by my recommendation, is the son of a market gardener, and knows nothing of botany. He has no appointment or salary, and means to settle there with a wife, as a farmer and market gardener.

"How you can be useful to your employers, as a botanical traveller, to send home seeds of plants from thence, till you have made yourself acquainted with those already in England, I don't know. We have now several hundreds of such, and to send them again would be idle and useless.

"You might discover some drug, valuable in dyeing or medicine, for your own advantage; but unless you are able to benefit your employers, as well as yourself, how can you expect employment? You are certainly very eminently capable of searching the woods, with diligence and advantage, for dyeing drugs, and other matters likely to be advantageous to manufactures and trade; and that many such things remain unknown in the unexplored wilds of a country larger than all Europe, is a matter of infinite probability.

"If the gentlemen of Manchester will make a subscription to maintain you in that employment, on such terms as shall be agreed upon between you and them, I will readily become a subscriber, and use my best influence with Government to send you out at the public expense, in which I have no doubt of being successful.

"I am, Sir,
"Your very humble servant,
"JOSEPH BANKS."

The plan of sending Caley out by subscription, as suggested by the right hon. baronet, if it was ever attempted, certainly met with no success, and he had the prospect before him of having to settle down to his old employment. The autumn passed without any realization of his hopes, or any apparent probability of their being realized, and it may be imagined that Caley's sanguine temperament received a severe check. About the end of November, however, to his great joy, Sir Joseph Banks hastily summoned him to London, in expectation of despatching him to the *terra incognita* he had so ardently longed to explore.

All that Caley had wanted and striven for was now promised. He was to be sent to New South Wales, and his primary duty was defined to be the collection of botanical specimens for his worthy patron, and of seeds for the Royal Garden at Kew, with liberty to preserve duplicates for his own use. No man could have been better adapted for such an undertaking. "He possessed a robust constitution, was a stranger to silken ease, and had a power of endurance and of suffering privation to a great degree. He knew how to conciliate the natives by an easy and jocular familiarity, and afforded at all times to his companions a fine example of perseverance. We have heard him say his spirits never yielded when he had to lead a party through the woods, though they had frequently done so when he was led by others; but these were only his physical qualifications."—'Magazine of Natural History,' vol. ii.

A voyage to Australia, in those days of slow sailing, must have been beyond description wearisome; but rarely, says Dr. W. Withering, has the tedium of such a voyage been more effectually or advantageously dispelled than by the varied studies which Caley, during the whole progress, unweariedly pursued. "We have seen," he adds, "the lone wanderer, irresistibly impelled by the contemplation of Nature in her grandest, yet most savage, form, penetrate the parched deserts of Africa, and, in search of all-captivating novelty—discarding the primary instincts of the mind—approach even the ruthless tiger's lair, as though unconscious of danger; but Caley was destined to less hazardous shores; it was his luck to be wafted to more temperate climes; and, while exploring the flowery prairies surrounding Botany Bay, instead of encountering the Mauritanian lion, he felt no fear but that of scaring away the timid kangaroo."

This employment was the summit of Caley's ambition. It was the utmost he had ever, in his most sanguine mood, hoped to attain. Ten long years, fruitful in botanical discoveries, and relieved by much adventure, were spent by him on the Australian continent. His faculty of observation was strong, and he seized, as if by instinct, upon the peculiarities of whatever met his eye. He added largely to the existing knowledge not only of botany, but also of zoology, during his residence in that remote quarter of the globe—making frequent, long, and sometimes perilous excursions into the bush for that purpose—and, what was of the utmost importance to his employers, he proved himself an extremely good preserver of specimens.

Mr. Brown, a celebrated botanist of that day, who, in company with Mr. Bauer, visited Australia during Caley's residence there, was greatly interested in his labours. '*Botanicus peritus et accuratus*'—

such was his opinion of the man; and as further testimony of the esteem in which he held him, equally with his patron, he named a newly-discovered orchidaceous plant (afterwards cultivated at Kew), *Banksia Caleyi.*

"How satisfactorily Mr. Caley justified the confidence placed in him," wrote his friend Dr. W. Withering, "is well known. Indeed it appears by his letters from Paramatta, Sydney, and other stations in the colony of New South Wales, that as the illimitable field of Nature expanded before his enraptured gaze, proportionally did his powers of observation become enlarged. No branch of natural history seems to have been neglected; and the extensive collection of quadrupeds, birds and reptiles purchased in 1818 by the Linnæan Society, and still [1830] constituting the most splendid portion of their museum, will remain a lasting monument of his successful efforts." Some account of this collection, brought over by Caley in the year 1811, was given in the 'Transactions of the Linnæan Society,' vol. xv., where ample justice is done to the collector's merits.

It is related of Caley that, having some claims to settle with the Treasury for his expenses home, he astonished the clerks in that department, who seemed not to have been familiar with such instances of ingenuous honesty, by refusing the amount offered him. He told them that it was not his intention to have returned at the time when he did (he was brought home to give evidence on behalf of Governor Bligh) and that he could not think of accepting more than, in his own judgment, was due to him.

The traveller had not been long in this country before he sought out his old botanical acquaintances, with some of whom he had corresponded during his absence; and the meetings which took place were joyous ones. The Middleton District Botanical Society was still in existence, and into its work he threw his whole energy, attending its meetings, joining in its excursions, and entertaining the members with stories of his adventures abroad. One of the most intimate friends of Caley at that time was Edward Hobson. These men were more than friends; they were as David and Jonathan. Their lives—so long as Fate did not decree their separation—were bound up together in fraternal union, for they were two of Nature's truest sons, and owned her irresistible spell. Old John Dewhurst, the president of the collective botanical meetings at Manchester, who had a great regard for Hobson, and possibly felt a little jealous of his newly-found friend—used to complain that Caley and Hobson were so much together that they could not spare time for conversation with anybody else.

After a few years' residence at home, Caley was offered by Government, and he accepted, an appointment as successor to Dr. Anderson, the superintendent of the Botanical Garden at St. Vincent. He found the expenditure of that establishment, upon his arrival in the island, greatly in excess of the requirements; and, prompted by a sense of honour, he devised measures of economy—for which, however, he got no thanks, either in the island or at home—and saved the Government some hundreds of pounds a year. This was effected by a reduction of the establishment to somewhat reasonable limits. He succeeded, by process of law, in recovering for the gardens a piece of ground which had been alienated during the time of his predecessor.

After a residence of eleven years in St. Vincent, which seems to have been anything but comfortable—making the total length of his residence abroad twenty-two years—Caley returned to England, and the establishment which he had superintended, and which had apparently served no useful purpose, was broken up.

During Caley's second residence abroad, a regular correspondence passed between him and Hobson; and the latter received from his friend many rare specimens of tropical plants, especially of ferns.

On his return to England, Caley—for what reason does not appear, for his friend Hobson, though infirm, was still living—settled at Bayswater, and led a somewhat secluded life, being supported by the scanty savings which he had been enabled to effect at St. Vincent. His wants, however, were few, for no man could have been more simple in his mode of living. His income is said to have scarcely exceeded that of an ordinary labourer; nevertheless he contrived to purchase, at various times, several hundred volumes of books (chiefly botanical books and records of travel), which were the solace of what would otherwise have been many a weary hour. He lived within his humble means, and honourably discharged every debt he owed in the world. The accounts which he read of voyages and travels enabled him to fight his own battles again, and recall his early struggles; and it was not difficult, he used to say, to place himself in the position of the hero in many a story of hair-breadth escapes. "He knew what it was to be hungered and athirst, to be drenched and to be naked, and to spend day after day face to face with death. Yet, after all, it was the mode of life he loved; and, if he had had his will, he would have returned to be a child of the woods again." Thus wrote one evidently well acquainted with his habits and disposition.

Though suffering from an impaired constitution, the consequence of his prolonged residence in the West Indies, Caley, whilst living in Bayswater, and finding his chief recreation in reading, did not forget his friends in Lancashire, or cease to feel an interest in their occupations. In his time no list of the plants found indigenous about Manchester had been published, nor did one appear for a long time after; but Caley, writing to Hobson on the 5th February, 1826, urged him to undertake the preparation of such a list, and said that he (Caley) prepared one—though it was incomplete—as far back as 1798.

This early list was not preserved. Caley informed his friend that it was not as copious in phenogamous plants as, from the general appearance of the country and the diversities of soil, might have been expected; and that some of the commonest plants in the kingdom might be reckoned amongst the scarcest in the Manchester flora. Many plants, he said, had become naturalised about Manchester that were not met with there formerly; and others, again, had become extinct. "John Dewhurst," he adds, "could give the best account on this subject; and it will be well to note down what he says."

Caley suffered acutely during the later years of his life from an obscure disease induced, as he believed, by an accident which befel him in St. Vincent. It was long before his friends could induce him to call in medical help. He relied upon his own skill, acquired under circumstances of necessity, in the treatment of ordinary ailments; and even when suffering from a disease which was calculated to baffle the most skilful physician, he abandoned his self-confidence regretfully. When visited by his friends, he took every opportunity—although often suffering excruciating pain—of conversing upon topics in which they felt a mutual interest; and even when bed-ridden, and unable to move without assistance, he set about correcting current errors respecting certain British plants.

The following is a recorded instance of one of these: Caley, in his rambles about Ingleborough, had gathered in abundance a *Hieracium* entirely new to him. He showed it to Dr. Withering, who was equally ignorant of it, and the two agreed to refer it to Dr. Dickson. By this gentleman the species was declared hastily, and on apparently insufficient grounds, to be *H. villosum*, and it was so recorded in the third edition of Withering's 'Arrangement,' although Caley averred that Dr. Dickson had but slender ground for his opinion. That he was wrong was shown by the plant turning out to be something else in the ensuing season. As far, therefore, as Withering is concerned, *Hieracium villosum* must be excluded from the British flora. Probably Caley's plant was *H. eximium*.

The manner in which Caley arranged his affairs when he felt the hand of death was upon him, is worthy of record. He had no large stock of worldly goods; but out of such as he had, his first desire was to provide for one who had tended him in his sickness; and his second was to repair, as far as he could, the injury he thought he had done to a poor bird which, caught in the wilds of Australia, had been his captive companion for twenty years. In consideration of this he charged in his will certain persons who were to be benefited by his property to tend and provide for all the wants of his cockatoo so long as it lived. To a negro slave in St. Vincent, who had no doubt found Caley ever a tender-hearted master, he bequeathed freedom; and he left the residue of his effects to his nearest relations. He had no family. His wife, whom he married in 1816, had died. "These are slight traits of character, but they mark a nobleness of mind which will for ever distinguish the possessor from the common herd of mankind. If poor Caley had had only the pocket of a beggar, he would have acted with the honour of a prince."—'Magazine of Natural History,' vol. ii.

Caley lived to see the spring of 1829—with what different feelings must he have regarded the opening year from those of his impulsive youth!—but as it advanced his strength declined. This declension was succeeded by an abatement of pain, until at length he was unable to converse, and on the 23rd May his death took place. His remains were deposited in the burial-ground belonging to St. George's Church, Hanover Square, beside those of another Australian traveller—Captain Flinders.

EDWARD HOBSON, OF MANCHESTER.

EDWARD HOBSON, a celebrated muscologist and correspondent of Sir W. Jackson Hooker, Mr. William Wilson, Dr. Greville, and others, was the pride of the old Lancashire botanists. He was for many years their recognised head—the infallible authority to whom disputed points were referred—and his premature death, in the autumn of 1830, about fifteen months after that of his friend George Caley, was to them a source of unspeakable grief.

The circumstances of Hobson's birth and education were as unpropitious as could well be conceived. He was a native of Manchester. When not more than three years of age he lost his father by death; and his mother having shortly after that event contracted habits of intemperance, he was placed under the care of an uncle, Mr. William Hobson, of Ashton-under-Lyne. He attended a school in that town, kept by a Mr. Wrigley, for what length of time is not known; and afterwards, whilst living with his grandfather in Manchester, he had a further course of schooling which terminated when he was about eleven years of age. This is believed to have been all the school education of which Hobson could boast.

What led him to undertake the study of botany—

whether, as in the case of George Caley, it was an early and spontaneous love for the subject, or whether he was infected by the passion for plant-hunting of those around him—is not clear. John Horsefield, one of his early associates, who survived him about twenty years, was of opinion that he received his earliest instruction in the science at a meeting of the Manchester Society of Botanists, of which he afterwards became an active and intelligent member. Horsefield first became acquainted with Hobson in the year 1809, and he thus relates the circumstance: "My father once found a plant of wild bugloss (*Lycopsis arvensis*), and neither he nor I could properly make it out. I took a specimen to one of the general meetings held at Radcliffe Bridge, where a young man from Manchester, to whom I applied for information, told me its correct name: that man was Edward Hobson. This was the first time that I recollect having met him, but ever afterwards, when I was in want of information on botanical subjects, my application was made to him for it; and in general he was both able and willing to give it to me."

John Dewhurst was president of the Society at the time spoken of, and Horsefield and his father had been members some years before Hobson's initiation. The probabilities are that Hobson, with the modesty which ever marked his character, had pursued alone, or with such assistance as he could privately obtain, his early botanical studies; and that it was only on his acquiring substantial knowledge that he ventured to associate himself with men who were veterans in the science. At all events, many years did not elapse before he was regarded as the most able botanist in Lancashire, and as such, on the retirement of Dewhurst, he was chosen to preside over the Society's meetings. Horsefield says: "Hobson attached the highest value to the acquirements of John Dewhurst, who for more than twenty-five years presided over these meetings. Being far advanced in life, Dewhurst at length resigned his post in favour of Hobson, whose more active habits better enabled him to keep pace with the advancing knowledge of the time."

Hobson turned his attention to cryptogamic botany, for the study of which his qualities of acuteness and steady perseverance peculiarly fitted him. It was in that direction that he became celebrated, and his attainments secured for him the warm attachment of some of the most eminent botanists then living. Horsefield related many instances of his daring attempts to climb trees and rocks in pursuit of rare mosses and lichens; and described some laughable disasters that befel him in detaching specimens from their resting-places. Ashworth Wood, near Rochdale; Baguley Moor, Cheshire—now enclosed and under cultivation; and Cotterill Clough, a romantic dell about ten miles south of Manchester, which has been a famous botanical ground for generations, were his chief places of resort; but he would make occasional excursions also to the hills around Staleybridge and Greenfield, and through the Derbyshire dales. "In fact," said Horsefield, "our excursions were varied in almost every direction around Manchester. We sometimes extended them to ten, twelve, and even twenty miles, always with a determination to return home the same evening."

To a person fond of natural history, and residing in the country, Hobson's society was invaluable. He appeared at all times quite as much gratified in communicating as in acquiring knowledge; and from his uncommon quickness and accuracy—such was the testimony of an old acquaintance—every wall in a garden, and every field, every lane, every brook or pond, afforded him opportunities of pointing out new or unobserved sources of gratification. When taking his favourite walks, the moment he found himself clear of the smoke of Manchester, his eye was on the alert in every direction, and his countenance, at all times pleasing, assumed peculiar animation whilst he was breathing the pure air of the country.

The late Mr. J. Moore, F.L.S., relates: "Not many years before his death he was so kind as to accompany me on an angling excursion to Bakewell, in Derbyshire, with the view also of obtaining something like an outline of the natural history of the River Wye. He was astonished and delighted with the endless variety of water-bred flies we met with, and especially by the many delicate specimens of the two great families Ephemeridæ and Phryganidæ, which appeared to have escaped the attention of our most careful entomologists. A better satisfied or more bustling trio has seldom been seen on the banks of that beautiful river than myself, battling with a good vigorous trout; an active little boy with my pannier on his back, twisting and turning his landing net in every direction to get the fish into it; and Hobson all the while in full pursuit after some newly-born ephemera across the meadows. During this visit we were quite satisfied that a large proportion of our Ephemeridæ and Phryganidæ are seldom seen, except by anglers; and had Hobson's life been spared, the acknowledged accuracy with which he had applied himself to the diminutive beauties of the vegetable kingdom would have been most willingly devoted to the splendid little insects which, in their short-lived existence, occasion to the disciples of Izaac Walton, as well as to the entomologist, an ever-varying interest in the matchless scenery of the Derbyshire rivers." Mr. Moore introduced Hobson to some intelligent friends staying at the 'Rutland Arms,' Bakewell, "who were, as might be expected, much pleased with his conversation and manners."

"It has been remarked," says the author of the short memoir from which this extract is taken, "that lovers of natural history live their pleasanter days many times over. It might be truly so said of

Hobson, for I believe a happier man is seldom seen than he was when engaged in arranging the insects or stretching out the mosses he had collected during his most successful rambles. With his imperfect instruction in ancient as well as modern languages, it is difficult to account for his being so well able to keep up with the new arrangements that were continually taking place in the different branches of natural history to which he was attached, and especially with the endless changes which occurred in descriptions of very abstruse derivation." Whenever he was so fortunate as to find, in the works of foreign authors, an engraving of any insect or plant he was studying, he had a sure resource in the then president of the Manchester Natural History Society, who most willingly translated the description for him; but Mr. Moore was not aware of any other aid that Hobson could reckon upon.

With regard to Hobson's attainments and the perfect honesty and simplicity of his character, ample testimony was borne by many who survived him, and who had been his never-failing friends. Shortly after his death Mr. Shaw, of Bollington, wrote to Mr. Moore: "Hobson first introduced himself to me sixteen years ago, by a visit to my little botanic garden, as a collector of specimens; and from the first interview our communications were made with that frank and open generosity which was so conspicuous in his character." Horsefield, again, writing to John Hampson, one of Hobson's most intelligent and valued associates, said: "Hobson was a profound practical muscologist, and never could have collected materials for his work had he not possessed the greatest patience and perseverance in his laborious investigations."

Amongst Hobson's acquirements drawing occupied a place, and Horsefield is said to have had in his possession a little book containing nearly 200 coloured drawings, exhibiting the generic and specific characters of mosses on a magnified scale, copied from a work in the Chetham Library, which he (Hobson) often visited during his dinner-hours.

Whilst thus extensively engaged in scientific pursuits, Hobson held a situation in the establishment of a Manchester manufacturer, Mr. Joseph Eveleigh—himself a naturalist and mineralogist of some local celebrity—and was a most conscientious servant. He was also an affectionate husband and father; and never permitted his fondness for science to interfere with his duty of providing for the daily wants of a large family. No incident could bring out his character more forcibly than the following, which may fairly be introduced as deserving of study by every man similarly circumstanced. It is told by Mr. Moore: "In the year 1829, having distinguished himself in assisting to arrange the museum of the Manchester Society for the Promotion of Natural History, it was unanimously resolved to offer him a permanent engagement in that institution; and Mr. Blackwall and myself were deputed to wait upon him for that purpose. Knowing his fondness for such pursuits we had no doubt that the situation, with a salary of £100 per annum, would be exactly what he would desire. His reply to the offer I must endeavour to give in his own words. Having recovered himself a little from the feelings which evidently overpowered him, he said: 'Gentlemen, I am deeply sensible of the great compliment and the kind attention paid to me by the offer you have made. The situation and the salary proposed would have been everything I could have wished for; but my present employer was very kind to me in his prosperity, and, in his altered circumstances, as I have every reason to believe my services are of more importance to him, I cannot think of leaving him.' "

He continued in Mr. Eveleigh's employment until a failing constitution compelled him to relinquish every occupation.

The extent of Hobson's service to botanical science is attested by his correspondence with the authors of the 'Muscologia Britannica' (Hooker and Taylor), by whom he was often named as one of their safest authorities. It was about the year 1815, when engaged in compiling his 'Musci Britannici' *—a copy

* 'The Musci Britannici: a collection of Specimens of British Mosses and Hepaticæ, systematically arranged with reference to the "Muscologia Britannica," "English Botany," and "British Jungermanniæ," &c. &c.' The copy referred to bears this inscription in the author's writing: "Presented to the Chetham Library by Edward Hobson, May 26, 1827."

of which is now in the Chetham Library, at Manchester—that his correspondence with Sir William Jackson Hooker (then Dr. Hooker) and other scientific men, commenced. He received material aid from them in the shape of rare specimens of mosses not procurable in his own neighbourhood, as appears from the correspondence.

Dr. Taylor, writing on the 10th of September, 1815, acknowledges the receipt of some valuable specimens, and says: "You will be surprised, my dear Sir, at my desiring to have so many specimens of those things which you find in your neighbourhood, and which appear to me rare; but the fact is that only thus can the science of botany be rapidly progressive, more, certainly, being to be learnt from specimens than from the very best plates with the very best descriptions. The winter and early spring are approaching—the season for mosses—when I trust you will favour me with some specimens."

In a letter dated the 24th of March, 1816, Dr. Hooker acknowledges the receipt from Hobson of some scarce mosses, and says that he is engaged in publishing a continuation of the 'Flora Londinensis,' with figures of every known species; and will be glad to receive specimens and any information respecting them with which Hobson can supply him.

A letter from Dr. Taylor, dated 11th of April, 1816, is highly encouraging to Hobson. "I was much pleased," the Doctor says, "with the mosses you were kind enough to send me; and if you will let me have another list of your wants I will endeavour to supply it."

On the 3rd of February, 1818, Mr. (afterwards Sir Charles) Lyell commenced a letter to Hobson thus:—

"DEAR SIR,—Our kind friend Dr. Hooker has begged that I will be the channel of conveying to you his admiration of your enthusiasm and acuteness in the study of British mosses, and his obligations to you for your remarks and every service in your power, by presenting you with his copy of the 'Muscologia' (which happened to be in my hands). I forward it to you with great pleasure; and have endeavoured to render the present more acceptable by the accompaniment of some Jungermanniæ and other cryptogamia of the New Forest."

On the 8th of May, 1818, Hobson received from Dr. Hooker a letter acknowledging the receipt of the first volume of his 'Mosses' in the following satisfactory terms:—"MY DEAR SIR,—Your packet I received yesterday, and am very much obliged to you for the copy of your 'Mosses.' They are very correctly named, and got up just as I could wish them." As this volume was illustrated with dried specimens of mosses and hepaticæ, instead of engravings, a few copies only could be furnished; and Hobson wrote to Mr. Scott, of Edinburgh (13th of June, 1818), to inform him that he had received his kind order for the first volume of 'Mosses and Hepaticæ,' with the £1 note enclosed, and that the other volume would be published as soon as sufficient material could be collected.

Dr. Greville, the distinguished author of the 'Flora Edinensis,' writing from Wyastone, near Ashbourne, 9th of August, 1819, thus addressed Hobson:—

"Dear Sir,—Since I had the pleasure of seeing you, I think I have been fortunate enough to discover a new species of *gymostomum*. I send you the only specimen I can spare. I shall be impatient till I hear from you."

Again, in a letter dated the 7th of February, 1821, Dr. Greville requests Hobson's assistance in procuring specimens of mosses—a list of which he sends him—and expresses regret that he cannot, in return, send Hobson all the specimens he wants for his second volume of the 'Musci Britannici.'

In September, 1821, Hobson, in a letter to Dr. Hooker, complains of having been so confined by his business that he had not had much time to devote to his favourite study; yet what time he had to spare he employed in laying down specimens for his second volume; but it is certain he cannot complete it without adding some of the lichens, unless something is done to procure for him the rarer species he wants; and he informs the Doctor that Mr. Eveleigh, the bearer of his letter, had a good collection of specimens of minerals as well as plants, and would convey any duplicates of rare mosses or jungermanniæ which the Doctor or his friends could furnish for his second volume, which he was anxious to complete as early as possible.

The following letter from Dr. Greville, dated the 8th of May, 1822, after the completion of the second volume of the 'Musci Britannici,' must have afforded Hobson much gratification:—

"MY DEAR SIR,—I beg to return you my best thanks for your second volume of 'Mosses,' in which I do not see anything that requires alteration, nor will Dr. Hooker, I think. I suppose you mean to proceed to a third volume, after you have made up your copies for the second. If you were to take in the Jungermanniæ and the lichens, you might go on for a good while; they would take much less trouble in preparing.

"Hooker thinks about a new edition of his 'Muscologia Britannica.' I am working very hard at my 'Flora Edinensis.'

"Yours very truly,
"R. GREVILLE."

Hobson had not leisure to complete a third volume of the 'Musci Britannici,' but the value of the work which he had already done is fully attested by this correspondence. We find him, however, a few years afterwards, cultivating an acquaintance with entomology; and Dr. Hooker, writing to him on the 1st of February, 1825, after observing that he had long been in his debt, informs Hobson that, having given up entomology for ten years, he regrets he is unable to render him any assistance in that pursuit. Jethro

Tinker, a correspondent of Hobson's, residing at Staleybridge (who died only two years ago), had in the preceding year furnished him with the names of the insects found in that neighbourhood; and called his particular attention to the circumstance of the very few butterflies to be met with. The pursuit of entomology, however, did not, as his correspondence with Caley and others about that time shows, efface his love for the science which had established his reputation.

In a leter to Mr. Henry Baines, of York, sub-curator of the Yorkshire Philosophical Society, and author of 'The Flora of Yorkshire,' 1840, Hobson says (6th of November, 1827) that he has been only a short time engaged in entomology, and has no collection of insects of which he can boast. He adds that botany has been his favourite pursuit at his leisure times, but that he then wished to combine both, "as it was no great additional burden to carry," and he hoped by a little diligence to do something in it. He also mentioned having sent to Mr. Baines an 'Entomological Nomenclature,' which he had got a friend to print for him, taken from Samouell's 'Compendium,' with a few additions which might be useful to him for cutting up to put to his collection, or might answer as a memorandum book, to know what he had got.

On the 27th of November, in the same year, Hobson wrote to Caley, then living at Bayswater, to inform him that in consequence of the extension of buildings around Manchester many of his favourite resorts were so altered as scarcely to be known, and that he had not been able to find a single specimen of a plant formerly growing in Scarweal Clough, of which Caley desired to be possessed, "seven or eight houses having been built upon the top of the bank, and the Clough cut up into gardens."

That Hobson's old friend had not, within less than a year of his death, lost his interest in Manchester botany is shown by a letter dated the 26th December, 1828, in which Caley asks Hobson if he has ever paid attention to the varieties of the bramble; and he mentions Baguley Moor, Sale Moor, Ashton Moss, and Sinderland Moss, as places where two very distinct species might be found, differing both in the form of the flowers and in the colour and shape of the fruit.

Hobson died on the 7th of September, 1830. His remains rest in the burial-ground attached to St. George's Church, Hulme, Manchester, where a mural tablet, erected to his memory, and bearing the following inscription, may still be seen:—

"SACRED TO THE MEMORY
OF
EDWARD HOBSON,
OF
MANCHESTER.
OBIIT 7TH SEPTEMBER, 1830. ÆTAT. 48.

"Humble parentage had afforded him only a scanty education—the necessary support of a numerous family demanded his daily labour.

"Yet amidst privations and difficulties he had, by assiduity and zeal, rendered himself a most skilful Naturalist, as his scientific works and ample collections lastingly testify.

"Entomology, Botany, and Mineralogy were his favourite studies: in these, many celebrated men, publicly in their writings, and privately in correspondence with him, have acknowledged his great attainments.

"Such distinctions did not affect his natural simplicity of manners; his character was wholly amiable."

The esteem in which Hobson was held by Sir W. J. Hooker is attested by the following letter to John Hampson, which forms perhaps the more gratifying memorial:—

"Glasgow, October 2nd, 1830.

"SIR,—I was much concerned and surprised to learn by your letter of the 17th of last month, and by the copy of the 'Manchester Guardian,' which you had the goodness to send me, that your friend and my valued correspondent, Mr. Hobson, had died. I was not even aware that he had been in an indifferent state. His loss will be severely felt by the lovers of British botany generally, for I hoped that, had he lived, it was his intention to have continued his 'Musci Britannici,' or rather to have extended the plan, so as to have included the whole of the British Cryptogamæ.

"I should be happy, were it in my power, to have furnished you with particulars relative to his general botanical knowledge and acquirements, but unfortunately nearly all I do know of him is by correspondence; and his modesty was such that he seemed to shun making anything like a display of his abilities and of the extent of his acquirements, and it was only incidentally that I discovered that he paid any attention to phænogamous plants. Such was his acuteness, however, and so completely had he mastered all the difficulties that attend the study of cryptogamic plants, that it was easy to perceive that the phænogamous tribes would have been comprehended by him with facility.

"It is, however, as a muscologist that Mr. Hobson's name will rank in the annals of botany. I do not know any naturalist who searched for mosses more successfully than he has done in their native stations; nor one who discriminated them more accurately.

"His publication of 'Specimens of British Mosses and Hepaticæ' will be a lasting testimony to his correctness and deep research into these beautiful families; and in this country he has been the first to set the example of giving to the world volumes which are devoted to the illustration of entire genera of cryptogamic plants, by beautifully preserved specimens themselves. This method has been pursued by Mr. Drummond in his 'Mosses of Scotland' and in his inestimable work on the 'American Mosses.'

"Once, and only once, I had the pleasure of a personal interview with Mr. Hobson. He came to me at the inn, in Manchester, bringing with him many of his new discoveries, and I scarcely knew which most to admire in him—his accurate knowledge of every plant he had investigated, or the extreme diffidence and modesty he displayed in communicating that knowledge. He had then in the examination of mosses only a common pocket lens to make use of; and I had the satisfaction of giving him my Ellis's aquatic microscope, by Jones, which had been my companion for many years, and which was the very last I ever employed.

"I have every reason to believe that this instrument opened to him new wonders in the vegetable creation, and contributed not a little to his very accurate knowledge of the minute cryptogamic vegetables.

"If you propose raising a subscription in the Botanical and Horticultural Society of Manchester, with the view of purchasing Mr. Hobson's collection of plants, for the use of that society, I shall be happy if you will set my name down for 5*l.*; and if you will let me know when the purchase is made, I will immediately remit the money.

"I am, &c.,

"W. J. Hooker."

The herbarium which Hobson had collected was purchased for the Manchester Botanical and Horticultural Society. His insects were presented to the Manchester Mechanics' Institution and formed part of the museum which, having been accumulated chiefly by the Banksians, for some time existed there. His friend Mr. Moore wrote of him: "I have reason to believe that the highest wages Hobson ever received were not more than 40*s.* per week, and that for many years they did not reach half that sum; yet he always kept himself out of debt, and, by the innocence of his habits and pursuits, secured to himself a portion of real happiness which is not often exceeded. In his anxious exertions to support his large family he afforded a most valuable example of integrity, punctuality, and diligence in the service of his employers, and made himself many friends. He had very early in life satisfied himself that in no way could he so agreeably or so safely recruit himself after labour, as in the quiet study of natural history; and this impression, added to his fondness for the science, occasioned a degree of perseverance which has seldom been equalled." The example of Hobson's life was not without its effect, for it led others similarly circumstanced to seek for relaxation and enjoyment in the same inexhaustible resources.

THE BANKSIAN SOCIETY OF MANCHESTER: HOBSON'S LAST DAYS.

Edward Hobson was mainly instrumental, in the beginning of 1829, in banding the local naturalists together in what was called, in compliment to Sir Joseph Banks, the "Banksian Society," respecting which it is necessary here to give some information.

The object of this society was "the acquisition of knowledge in the sciences of botany, entomology, mineralogy, geology, &c.," to be effected "by means of a library of books, conversational discussions, the reading of papers, and occasional lectures, illustrated by collections of specimens in the several sciences."

The minute-book of the society, now in the possession of Mr. Robert Crozier, of Manchester, an artist of considerable repute, whose father was a worthy Banksian, shows that it was established at a meeting held on the 5th January, 1829, at which Mr. Rowland Detrozier presided. After the adoption of a code of rules the officers were chosen by ballot, as follows:—President, Edward Hobson; treasurer, William Garry; secretary, John Hampson. Committee: Messrs. Rowland Detrozier, William Hobson, Thomas Hewitt, John Longbottom, Thomas Forster.

It was resolved that for the formation of collections each member should contribute annually at least three perfect specimens; and the following plan of arrangement was to be observed:—

"ENTOMOLOGY.—The insects to be arranged according to the most approved system; and, in collecting British specimens, it is expected that members will, as often as practicable, observe their localities, habits, times of appearance, &c., that memoranda may be kept of everything interesting to the naturalist or useful to the Society. The foreign insects to be kept carefully distinct from the British, and labelled 'Foreign.'

"BOTANY.—A herbarium of the plants of Great Britain, and a distinct one for a general collection. Specimens of seeds, gums, sections of woods, or of any other miscellaneous articles, fossil as well as recent, which may tend to illustrate the science. As often as possible, the habitat of each specimen, distinctly written, and the time at which it was gathered, must accompany it.

"MINERALOGY AND GEOLOGY.—The minerals and geological specimens to be methodically arranged, and, as often as practicable, labelled with their names, and the situations in which they were collected."

At the inaugural meeting Mr. Detrozier, the chairman *pro tem.*, delivered an address, in which, after describing the objects sought to be attained, he pointed out the motivės which ought to stimulate the members to exertion. "Living at a time," he said, "when truth is no longer shunned, except by those who are interested in error, and when the progress of learning is no longer dreaded, save by those who profit by the ignorance of the multitude, whatever useful knowledge we may gain by mutual instruction—whatever new or interesting facts we may be so happy as to discover—the fate of a Galileo need not deter us from communicating to each other, nor from publishing to the world. Happily for mankind the consideration of what is true, in everything which relates to the existence, prosperity, and mental culture of man, is becoming more and more paramount to the consideration of that which is merely expedient; and the liberal and enlightened portion of the rich and more fortunate class of mankind contemplate the rising talent and increasing knowledge of the industrious artisan, not merely with complacency, but with pleasure. Still more happy, however, will it be for this latter class—which constitutes by far the greater portion of the human race—when the redeeming influence of knowledge shall be more largely felt and more duly appreciated amongst them, and when their leisure hours shall be consecrated to the attainment and communication of useful knowledge. Nor can I conceive anything more calculated to secure the realisation of this happy state of society than the establishment of institutions like the present. Time has led to the acknowledgment of the gratifying fact that, in the empire of mind, wisdom constitutes the only true riches; and experience, that best of all teachers, has confirmed the sister truth that industrious poverty may attain to its possession. And how gratifying is the fact to those who have little else beside their attainments of which to be proud, that the possesion of a full measure of talent is not confined to any particular class of society; that wisdom is not hereditary; nor yet like the unprincipled sycophant, found exclusively in the train of the rich and powerful." Noble sentiments! Would that the industrial classes of the present generation would take them well to heart.

"Brief as this sketch has been," concluded Detrozier, "it has, I trust, been sufficient to convince those to whom conviction was necessary, that the subjects contemplated by this society, as the basis of its labours, are worthy of their attention. The utility of such studies no man will doubt who is conversant with his own nature; for so long as a man is possessed of mental energies, those energies will exert themselves on subjects either prejudicial or favourable to happiness: and as relaxation is necessary to the wearied frame, that cannot be useless which combines interest and instruction with bodily ease. And even in the limited view which these sciences exhibit to the philosophic mind, it may catch from them a glimpse of the general economy of Nature; and, like the mariner cast upon an unknown shore, who rejoiced when he saw the print of a human foot upon the sand, it may be led to cry out with rapture, 'A deity dwells here!'"

During the existence of the Banksian Society (it was dissolved September 5th, 1836), the members accumulated botanical and entomological collections of considerable value, besides books. The balance-sheet at the end of the first year showed that they received—chiefly in the form of subscriptions—and expended, the sum of eighteen pounds. Edward Hobson's presidency of the Banksians lasted scarcely two years. A record of his death, accompanied by an expression of sorrow on the part of the members, and a brief account of their late president's career, was entered in the society's minute-book. The state of his health had prevented Hobson from following his usual employment during the spring and summer; and acting on the advice of his friends, he had resided at Bowdon. "It was only necessary," says Mr. Moore, "to make known to the respectable families in that neighbourhood that this amiable and interesting individual was sojourning near them in search of health, to secure for him every comfort which they had it in their power to furnish." Mr. Moore's last interview with Hobson was affecting. "His appearance indicated the near approach of death; and his countenance, always bespeaking benevolence to others, became expressive of the deepest gratitude whilst he pointed to the rare fruits and delicacies which had been sent to him by persons unknown. His perfect simplicity made him quite unable to account for such seasonable attention to a stranger. I promised that I would seek out his benefactors and thank them for him."

The regard for their president which was entertained by the Banksians is shown by the following entry in their minute-book, after a notice of the interment:—"The members of the society attended the mournful ceremony, and were joined by a number of botanists from the surrounding districts, who, on hearing of the sad event, spontaneously came forward to pay their last respects to departed worth and genius."

The death of their president was a source of great trouble to the Banksians. On the 21st September, when the first regular meeting after that event was held, no lecture was delivered, and it was deemed a matter of paramount importance "to consider of an eligible person to fill the vacant chair;" and the voice of the meeting was unanimously in favour of Mr. Peter Barrow. That gentleman, however, felt compelled to decline the appointment, but not without strong appeals on the part of the members. As showing the interest taken in the Banksian Society by men of rank, a letter is quoted from Sir J. W. S. Gardiner, strongly urging the necessity of making a judicious choice of a person to succeed Mr. Hobson in the presidency; and recommending Mr. Barrow, in whom were combined all the qualities necessary for such a post. The communication was well received, and it was resolved—"That this society entertains a lively sense of gratitude to Sir J. W. S. Gardiner, Bart., for his kind solicitude for its prosperity and for his excellent and proved advice in making a judicious choice of a person to succeed its late honoured president."

Mr. Barrow, having been elected by the unanimous voice of the members, yielded to the urgent solicitation of his friends, and accepted the office of president; but it was an office which he did not long hold. He was succceeded by Mr. Joseph Eveleigh.

A variety of causes contributed to bring about the dissolution of the Banksian Society, which it is not necessary, at this distance of time, to relate. At a general meeting of the members on the 5th September 1836—" It was unanimously agreed that the Banksian Natural History Society be now dissolved, and that its property be transferred to the Mechanics' Institution." An inventory of the property so transferred (some years afterwards, to the regret of many, treated very much as useless lumber) was prepared; and it included, amongst other things, insects and dried plants, collected with great pains and carefully preserved by the Banksians, a glass case of stuffed birds, minerals, furniture &c. The Society's books were placed upon the shelves of the Mechanics' Institution, and were, and still are, much valued by the members.

The dissolution of the Banksian Society did not, happily, put a stop to natural history pursuits in Manchester. Many of its members promoted the formation of a natural history class at the Mechanics' Institution, which was fostered by the directors, and of which the leading members were George Crozier, Samuel Carter, Charles Cumber, and Henry Day—all men of considerable attainments.

JOHN HORSEFIELD, OF PRESTWICH.

JOHN HORSEFIELD, although less distinguished than either Caley or Hobson, was nevertheless an accomplished botanist. His taste for botany seems to have been hereditary, for his father and grandfather both studied and cultivated plants—in a humble way, truly, but with a sense of the happiness which such an occupation ever gives. Referring to his ancestors, Horsefield wrote: "My grandfather was, I am informed, a good scholar—that is, he could read and write—and he was fond of botany, and fonder still of floriculture. He lived 84 years." And of his father he wrote: "My father could never write, and could scarcely read; yet he possessed a considerable stock of information, and was particularly fond of plants and flowers." It will he concluded from this that John's opportunities of learning were extremely meagre. When six or seven years of age he was sent to a weaver—who had perhaps, like his own grandfather, the accomplishments of a "good scholar"—to be taught reading. This weaver followed his regular employment at the hand-loom whilst his scholars, numbering at times half-a-dozen, sat by his side and read their lessons to him singly. His terms were two shillings per quarter! This was a kind of school now happily extinct.

Horsefield continued under the weaver's tuition for a period of twelve months. That he possessed some native genius is evident, for he says: "As soon as I had learnt to read—being set down to labour at home—I became immoderately fond of books, and read indiscriminately all that I could lay my hands on, upon every conceivable subject. A melancholy and rather fretful disposition, a feeble and delicate constitution, a sickly appetite, and a partiality for books, were the principal characteristics of my boyhood; writing and arithmetic were gradually added to my other accomplishments." His youthful aspirations in the botanical line were not greatly fostered by his parents, if we are to judge by the following passage: "I had noticed a few wild plants that grew in the fields about where I lived, and had applied to my parents to know the names of them, and what they were good for; but, not receiving what I considered a satisfactory account, I concluded that wild plants were not of sufficient importance to mankind to have names assigned to them. How different was the conclusion I came to on reading 'Culpepper's Herbal.' The wonderful properties that are there ascribed to plants excited in me a strong desire to get acquainted with the plants themselves." Certain it is that the interest excited by his reading of Culpepper remained with Horsefield, and in course of time he was enabled to procure more trustworthy botanical guides. If they are more commonly the property of the herbalist and the curioso rather than of the scientific botanist, still the 'Herbals' of Gerarde, Parkinson, and Culpepper deserve to be thought very kindly of for the stimulus they often give to the botanical inclinations of youth. Of Gerade's book, Dr. Pulteney, in his 'Historical and Biographical Sketches,' says: "It is a work which maintained its credit and esteem for more than a century; and pleasing as it is to reflect on the rapid progress and improvement of botany within the last half-century, yet there are many now living (1790) who can recollect that when they were young in science there was no better scource of botanical intelligence in the English tongue than the 'Herbals' from Gerarde and Parkinson."

Of systematic botany Horsefield long remained ignorant; following the occupation of a weaver from morning till night—being without books and without instructors—he had literally no means of obtaining knowledge. By-and-by, however, light dawned upon him. He heard the word "class," as applied to plants, first mentioned by a friend of his father, who showed him a species of Veronica, pointed out its two stamens, and said that these indicated the class in the Linnæan system to which the plant belonged. Here was something which excited the curiosity of the tyro. About the time in question—Horsefield being about 16 years of age—a botanical society of working men was formed at Whitefield, which his father joined. The members subscribed for the purchase of botanical works. "The first book that my father brought home," says Horsefield, "belonging to the society, was 'Lee's Introduction to Botany.' But what ideas of plants did I acquire by its means? Unacquainted with either Greek or Latin, and but partially acquainted with the etymology of the English language, hundreds of terms I found here for the first time; and, though it is more than forty years ago"—he wrote in 1850 —"I distinctly recollect the determination that actuated me to overcome the difficulties that lay in the way of learning them. The names of the 24 classes [Linnæan] were each formed of two words said to be derived from the Greek language, of which explanations were given. I wrote these 24 names down on a sheet of paper, and fixed it to my loom-post, so that when seated at my work I could always have opportunities of looking it over."

We have heard of a lady who acquired a fair knowledge of Latin during the moments stolen from her daily toilette; but a hand-loom weaver driving the shuttle and striving to commit to memory the classes and orders of the Linnæan system must have been a spectacle of no common interest.

"When in this manner," he continues, "by much perseverence and industry, the names and characters of the classes, as well as the orders, were well understood, and treasured up in my memory, every time that I went out into the fields and gardens, flowers were gathered and pulled to pieces, and referred to their proper place in the system—at least, such as I could so refer without any particular difficulty; but I was often very much puzzled." No doubt; and who can fail to sympathise with him?

In the year 1808, or 1809, he was introduced by his father to a general botanical meeting, held at Ringley Bridge, near Manchester, and from that time, young as he was, Horsefield began to associate with practical botanists. Hobson had not then joined the society. Its president and "namer of specimens" was John Dewhurst. The office of president of the collective meetings was one of no small importance. It was necessary that he should be an accomplished botanist—one able to name any specimen at sight, and never at a loss for a botanical definition—for it was the practice, at these general meetings, when a member had gathered a plant of which he required the name, to hand it to the president; and the president, after supplying the name and some information respecting it, would pass the specimen round for examination. Horsefield says: "At a meeting of the Middleton Botanical Society in April 1872, I met with George Caley, who had lately returned from a botanical excursion in New South Wales; and I was much amused with his account of his adventures, and the plants and animals he had met with. In the month of July following I went to meet Mr. Hobson at Four-lane Ends, Hulton, at a general meeting; and being selected to name the specimens of plants for the company assembled, he requested me to sit beside him while he performed that duty, and assist him if necessary; showing—Horsefield observes with pardonable pride—"that he had some confidence in my knowledge of plants at that time, though he needed little of my assistance."

The meetings at the time here referred to were little if at all affected by the excited state of political feelings which then prevailed. This is the more extraordinary because societies of higher pretensions—notably the Liverpool Philosophical Society—found it expedient for a time to close their doors. The botanists' meetings attended by Horsefield and Hobson, moreover, were located in the most excited part of the manufacturing district, for it was at Middleton, in the year 1812, that a most serious riot, in which a number of weavers were shot, occurred.

Who cannot sympathise with Horsefield also in the following pleasant reminiscence? In August 1812, "I went to another meeting at Seal Lane, Tyldesley, accompanied by a young woman, as well as my sister and her husband. It was a long journey, but the longer the better, with the company I had. Hobson was not there; so that I was requested to name the specimens. The young woman named, and I, young no longer, live together yet"—1850—"for we were married towards the close of that year."

There was a bond of cordiality running through

the society of which Horsefield was a member,—a willingness to render mutual assistance which is characteristic of naturalists; and Horsefield partook of some of its happiest features. As in his younger days he received help from those more expert than himself, so ever afterwards he was glad to help any rising genius whom he met. How he struggled, when a youth, to master the hard technicalities of botany shall be told in his own language. The society of which his father was a member did not exist for more than a few years, for want of a person competent to carry it on, and give the required information to members; and he says: "The only books that he had ever brought home for my perusal, as far as I can remember, were the one already mentioned (Lee's 'Introduction'), Withering's 'Arrangement of British Plants,' fourth edition, and Young's 'Latin and English Dictionary.' After the dissolution of the society, the books became dispersed amongst its late members, and for a while afterwards I could scarcely get hold of a botanical work of any description to consult, though I continued to attend the general meetings. But, in a year or two—1811 or 1812—James Shaw, of Prestwich, a friend of my father's, offered to lend me his share of books belonging to the Middleton Botanical Society, of which he was a member, provided I would return them to the society at one of its meetings—an offer which was cheerfully accepted by me—and in this way I got hold of two volumes of Turton's 'System of Nature,' which treated of the vegetable kingdom. *Out of these I copied* 1640 *essential generic characters for my own use;*" and this manuscript was not without its value thirty years afterwards, when he had access to scores of volumes. He had previously copied the generic characters and a list of the species of British plants from Withering's 'Arrangement.' The two manuscripts together served his purpose in the absence of printed books for a long time.

Mr. Leo H. Grindon, who was personally acquainted with Horsefield, refers to him in 'Manchester Walks and Wild Flowers' as one of the most celebrated of the Lancashire operative botanists. With the valley of the Irwell, and the adjacent cloughs and woods of Stand and Prestwich, his memory is peculiarly associated. "It was Horsefield," says Mr. Grindon, "who first showed me the way through Mere Clough, and pointed out the spots occupied by its rare plants. For thirty-two years he was president and chief stay of the Prestwich Botanical Society; and from 1830 up to the time of his death president also of the united societies of the whole district." Horsefield followed his occupation—that of a weaver—in a cottage at Besses-o'-th'-Barn, a village between Whitefield and Prestwich. Referring to his character and attainments, Mr. Grindon says that nothing could show more strikingly how an indomitable will and ardent thirst for knowledge, and a deep and faithful love of

Nature will triumph over the obstacles of poor means and humble station in life, and lift a man into the high places of true science, and give him at once the power of true usefulness to his fellow-creatures and of realising the true rewards of existence.

This estimate is justified when it is considered that Horsefield not only paid attention to botany, but also studied—in a very humble way, certainly, but very perseveringly—astronomy, algebra and mensuration; and found time, also, to take part in the political movements of his day, and to write verses. That an illiterate weaver should be dabbling in astronomical science was more than his neighbours could comprehend, or affect to be pleased with. He relates that when engaged in the construction of an orrery he applied to one Gideon Coop, of Prestwich, to make him a gilded ball to represent the sun. "Aw'll mak' thee one," said Gideon in the vernacular of the county, "an' charge thee now't for't; but, let me tell thee, fancy folk like thee 're a'lus poor" —"an observation," said Horsefield, "which my after experience proved to be correct; though I thought little of it at the time." The ball was made, the orrery constructed; but, alas! where praise and admiration were looked for, the outcome of Horsefield's ingenuity was regarded only as "a piece of lumbering nonsense." Eventually, unable to bear the jeers of his unscientific neighbours, the apparatus so skilfully constructed was laid aside: the planets, one by one, were lost, or stolen by his children for playthings; and "that thingumbob," as the children delighted to call it, which, wherever put, seemed to be in the way, was in course of time entirely broken up.

Horsefield was a Banksian, but he rarely attended the Mechanics' Institution meetings after the dissolution of the society, as he reserved himself for those country meetings where his knowledge and good nature had the full wide scope which (on the testimony of Mr. Grindon), they at once merited and deserved. "In person he was thin and spare, presenting a great contrast to the tall and patriarchal figure of Crozier, partaking, however, so far as we had opportunities of judging, of all his amiable, unsophisticated qualities." — ('Walks and Wild Flowers,' p. 58.)

Horsefield died on the 6th of March, 1854, having attained a good old age. No man ever lived whose moral feelings can be said more truly to have been—

> "Strengthen'd and brac'd by breathing in content
> The keen, the wholesome air of poverty."

His remains were followed to the grave by many of his old botanical companions, who, while he lived, knew how to appreciate the kindliness, the sterling honesty, and faithfulness to truth and right principle which characterised all his actions.

JAMES CROWTHER, OF MANCHESTER.

James Crowther occupied a station in life similar to that of Horsefield, whom he resembled also in his personal character and scientific attainments. That his educational opportunities were of the slightest description may be judged from the circumstances of his birth and parentage. He was born in a cellar in Deansgate, Manchester, June 24th, 1768, and was the youngest of seven children, his father being a poor labourer. When he had reached the age of six, his parents sent him to school—no doubt either a charity school or one of the class which Horsefield attended—and at nine employment was found for him as a "draw-boy" at petticoat weaving. Such education as he received was of little or no advantage to him in after-life. But he had a keen eye to everything around him, and natural objects—plants and insects—early excited his wonder, and prompted him to collect and study. Like others who in after life became celebrated, he had no books, no means of ascertaining the names of plants—except such common names as were prevalent in his locality—and it was long before the real starting-point for him of definite botanical study arrived. As in the cases of Caley and Horsefield, some incident—as the finding of an old Herbal, or the formation of a botanical acquaintance—furnished the requisite stimulus, so also in that of Crowther, the scientific description of a particular plant by a friend who had made some progress in botany, and the direction given to sources of further information, set him on the right track. He became an ardent botanist, joined the society of which John Dewhurst was president—which consisted entirely of working men, and met weekly during summer—and formed acquaintances which proved of great value to him. Dewhurst and Hobson became his closest friends, and their friendship, which partook of the sweetness of Nature, lasted throughout life. None of the vicissitudes of fortune, for thirty long years or more, ever sundered them.

It was characteristic of all the operative botanists in Lancashire that in the pursuit of their favourite science no fatigue, nor even danger, ever daunted them. If a plant grew within twenty miles of home of which they happened not to possess specimens, specimens must be procured at all hazards. If a day could not be spared—and those men could ill afford the loss of a day's wage—night must be devoted to the journey. Crowther was one of the most ardent spirits in the society of which Dewhurst held the presidency. Often, after completing a day's work, would he walk fifteen or twenty miles in quest of a coveted plant of whose habitat he had been informed. He usually reached his destination at daybreak, before other people were astir, and, having run the risk of being apprehended for a poacher or a thief, returned to Manchester in time for his next day's work.

He was not always fortunate enough to escape suspicion, nor even apprehension. Led sometimes by his perhaps too zealous search into trespassing upon game preserves, he was often pursued by the keepers; but, except in one or two instances, always escaped capture. He knew that if taken before the justices on a charge of trespassing in pursuit of game, the plea that he was merely botanising would have been a sorry one; so he relied, and usually with success, upon his running powers for escape.

One of his adventures is thus related: He was botanising in Hopwood Park, Lancashire, when the game watchers saw him and gave chase. Straight across country, for three or four miles, the pursuit was continued—it was a game of hare and hounds—but Crowther was the stoutest in wind and limb, and got clear off. He was not so fortunate in every instance. Once whilst searching for the cloud-berry (*Rubus chamemorus*) at Staley Brushes, he was seized; and it was with extreme difficulty that he persuaded Lord Stamford's gamekeepers that he had no designs upon the grouse. On another occasion when botanising on the estate of Mr. Egerton (father of the present Lord Egerton) of Tatton, he got into a difficulty which, had Mr. Egerton been a less considerate gentleman, might have had a serious termination. It was his habit, when in search of aquatic plants, as on the occasion in question, to carry with him a rod, not unlike a fishing rod in general appearance, having joints with brass ferrules, and at one end a pair of hooks. One hook had its inner edge sharpened, and was sickle like, its object being to cut off plants growing in the water out of arm's reach; whilst the other, which was not so sharpened, was used to angle the plants so detached to the bank. When engaged in this occupation on the mere side at Tatton, two gamekeepers came up and seized him; and notwithstanding all his protestations to the contrary, and his assurances that he was not angling for fish, but for plants, he was taken before Mr. Egerton on a charge of poaching. To that gentleman Crowther told his simple tale, exhibited his tackle and hooks, and showed clearly that he was not after the fish. Mr. Egerton, who saw at a glance that the implement was not adapted for fish poaching, directed that its owner should be immediately liberated; and as sometimes out of evil good will come, so in this case Crowther, as a reward for his brief imprisonment, had the satisfaction of hearing the keepers told that he (Crowther) had permission to go wherever he chose on the Tatton estate, and that he was not to be molested.

The excursions of Crowther and his companions frequently extended into Yorkshire. A favourite botanising ground with him, and one much valued at the present day, was the limestone district of Craven, where Malham, with its "cove" and subterranean rivulet, and Gordale Scars, rich in botanical rarities, are places of extreme interest. On those journeys he invariably gathered specimens of the lady's slipper (*Cypripedium calceolus*), a plant which —as its habitat became known to north country botanists—was in course of time exterminated. It is doubtful whether it is now found in any part of Yorkshire. It does not occur in Mr. Miall's list of Malham plants ('Naturalist,' vol. i.), a pretty sure indication that it is now extinct in that locality. All that Hooker and Arnot, in their 'British Flora,' can say of it is: "Woods in the North of England; very rare." One only known locality for the plant— the only one, at all events, recorded—is Castle Eden Dene, County Durham.—('Science Gossip,' vol. viii.) In the latter locality specimens were found by Mr. Henry Baines, the author of the 'Flora of Yorkshire' (1840), as far back as the year 1837.

In company with Dewhurst, Hobson, and others, Crowther made frequent excursions to Cotterill, Ashworth, Birtle, and Marple Cloughs—places all famous for their cryptogamic treasures—and also to the hills in the neighbourhood of Greenfield and Saddleworth.

The happiness which a new discovery always gives to a naturalist was well illustrated in the case of Crowther. His finding of *Limosella aquatica* on the borders of Mere mere, Cheshire, passed into a proverb, and was never reverted to but with delight. He was botanising with John Dewhurst when he made that discovery, and threw up his hat for joy.

On Dewhurst turning to see what fit had seized his friend, Crowther cried out that he had found a new plant, which was a perfect gem. Dewhurst shared his satisfaction at the discovery.

Hobson, on their return home, was informed of the circumstance, but professed himself somewhat sceptical, and said that unless he actually saw the plant *in situ* at Mere, he would not believe that it was a new discovery. Arrangements were made for a second journey, for the special purpose of gathering *Limosella aquatica*—Hobson to be of the company—and what happened is thus amusingly told by Mr. J. Moore, F.L.S., Hobson's friend :—" Hobson had great doubt as to their meeting with the desired plant, and when they were within sight of the lake, poor Crowther—whose accuracy was in question—had the mortification to find it so swollen with recent rains that the plant was at least three feet under water. Hobson felt for Crowther's disappointment, and set about botanising in the adjoining fields rather than complain of a fruitless journey. Whilst so engaged he heard a plunge in the water, and looking round, Crowther had disappeared. In the greatest alarm Hobson rushed back, and had the satisfaction to see the old man just emerge from the water with the precious specimen in his grasp."

Of Crowther's connection with some of the eminent botanists of his day, fragments of information remain. He rendered great assistance to Dr. Hull when that gentleman was engaged in the preparation of his 'British Flora,' by collecting for him rare plants within a considerable radius of Manchester. Among the upper class in Manchester he had many friends. Dr. Tomlinson presented him with a copy of Withering's 'Arrangement of British Plants,' which he much valued. For collecting a rare plant growing near Middlewich, Cheshire, a Mr. Carmeletti is said to have given him four shillings and a pair of new boots; one of the most munificent rewards, Crowther used to say, in his old age, that he ever received.

Crowther never allowed his wages (from sixteen to twenty shillings per week) to be trenched upon for the purpose of procuring a botanical book, or anything else in the shape of recreation. No man was ever more honest towards his family: he gave his wife every penny of his week's earnings. When he did want a supply of pocket-money—and he never required much—his *modus operandi* was as follows: After his day's work was done, he would go down to Knott Mill, the landing-place for passengers arriving by the Duke of Bridgewater's packet from Liverpool and intermediate places, and earn a shilling or two by the porterage of luggage. (There was at that time, of course, no line of railway into Manchester.) By this means all his requirements in the pursuit of his favourite studies were met.

The following incident is related, as showing the extent of Crowther's botanical attainments:—When the late Sir J. E. Smith was engaged on one of his botanical works, he spent some days with a Mr. Roscoe, at Liverpool. Happening to mention to his host that the publication of his book was delayed for want of information respecting certain mosses and lichens, Mr. Roscoe suggested that he should enquire of the weavers of Manchester, some of whom, he said, were good botanists. Sir James, who had not heard of these humble men of science, at first ridiculed the idea; but on being assured that he was likely to obtain from them all the information he wanted, he proceeded to Manchester by the Duke of Bridgewater's packet. At Knott Mill, Crowther was as usual plying for hire, and Sir James, without knowing the character of the man, engaged him to carry his luggage up to a certain hotel in the town.

On the way Sir James asked if Crowther knew a gentleman (whom he named) who lived at Hullard Hall.

"Very well," said Crowther; "he's a bit in my way."

"What way is that?"

"He is fond of botany, and a collector of mosses and lichens."

This led to further enquiry, and Sir James Edward Smith, as he afterwards declared, was furnished by his porter with all the information he was in search of. Crowther, it is said, received a shilling for his job, and another shilling with which to drink "Success to botany."

Crowther's general character amongst his fellow-students was high. He was a man of happy disposition, and the life and soul of any botanical party. Botany was not, however, his only study, although it was that to which he principally applied himself, and he excelled in the neatness and order with which he arranged his plants; he had some acquaintance also with entomology, like most of his early companions.

With Mr. Aiton, the predecessor of Sir William Jackson Hooker in the curatorship of Kew—a distinguished botanist, and author of the 'Hortus Kewensis'—he had a meeting similar to that which is recorded to have taken place between him and Sir James Edward Smith.

One evening, at Knott Mill, he was employed by a gentleman to carry some luggage into town, and conversation *en route* elicited the fact—which to the stranger was a matter of some surprise—that Crowther was a botanist. The gentleman (who proved to be Mr. Aiton) asked Crowther if he knew any distinguished botanist in Manchester. He was able to name two then living, Mr. Lee Philips (afterwards colonel of the local Volunteers) and Mr. Shepherd, who became curator of the Botanical Garden at Liverpool. Crowther in the course of conversation, mentioned as his most recent discoveries *Polypodium oreopteris* and *P. phegopteris*—ferns which were in his estimation extremely rare, as no doubt they were, and still are, in South Lancashire—and finally directed the stranger to the house of Mr. Shepherd.

Poor Crowther used afterwards to mention an incident which was a source of great bitterness to him. He was in Mr. Shepherd's house the day after his interview with Mr. Aiton, and overheard Mr. Aiton, who was in an adjoining room and ignorant of Crowther's proximity, express a desire to see again the man who had the day previously carried his luggage,—mentioning him by name. To his surprise and disappointment, Mr. Shepherd, though he knew of Crowther's presence in the house, evaded the inquiry; and Mr. Aiton left without seeing him—indeed he never saw him again. Shortly afterwards Mr. Aiton was asked to recommend a curator for the Liverpool Botanical Gardens, and he named Mr. Shepherd, who was appointed, and who filled the office until the time of his death.

In his younger days, Crowther had a strong sense of humour, and was fond of a good practical joke, as the following anecdote which is related of him will show:—Among the students of the Manchester New College (afterwards removed to York) was a young man named Roscoe, a member of the distinguished family of that name in Liverpool, who was an ardent botanist and friend of Crowther, whom he not only took frequent excursions with, but also employed to collect. On one of their excursions young Roscoe was genteelly attired in the costume of the period, that is "shorts," and white silk stockings, whose beauty Crowther was irresistibly tempted to spoil. Leaving his friend botanising on safe ground—(they were on the margin of a bog near Crumpsall)—Crowther went over the bog, picking his way from one clump of grass to another, until a particularly dirty patch, partially disguised by grass and *Sphagnum* separated them. Suddenly, with that demonstration of joy which always accompanied one of his discoveries, Crowther called to his companion to "Come and see a rare plant." Roscoe rushed towards the spot, but had not got more than half-way when he sank to the knees in soft mud, with which, on emerging, he found his delicate silk stockings thickly coated.

This was rather too serious a joke, but the subject of it bore his mishap with great good humour, and after partaking of some refreshment at a neighbouring public house, where he also got cleansed, he and Crowther walked home together. The offender did not go unpunished. Shortly after this trick was played upon him, Roscoe was visited at his lodgings by Crowther, whom he induced to take hold of the chain of an electrical machine, of the properties of which, of course, the poor man was ignorant—whilst the owner professed to be engaged in some delicate manipulation. The result may be imagined; Crowther got a stunning shock.

"There!" said Roscoe, "you bogged me; now we're quits."

In his declining years Crowther experienced, in an extreme degree, the cares and anxieties attendant on a life of poverty. He was a man of unblemished character—"one of the most simple-hearted men that ever lived," as we are told; "not *learned* but very *loving*." For some four years before his death (which occurred on the 6th January 1847) he had received a small pittance, 3*s*. weekly, from a society established in Manchester for the relief of scientific men in humble life. He had little reason, however, to thank that society for its help; for prior to its formation he had been in receipt of 5*s*. weekly from a few benevolent individuals, who, on his accepting aid from the society, discontinued their bounty. A sum of 7*l*. was subscribed to pay the expenses of his funeral and of a gravestone; and we are assured that one-seventh of that amount would have been a real blessing to the old man, during the last twelve months of his life, in procuring him warmer clothes and bedding. He had, under pressure of circumstances, sold the greater part of his botanical and entomological collections.

Poor Crowther's death is supposed to have been accelerated by the lack of proper clothing and nourishment. A married daughter—herself suffering many privations—had been his principal support. In acordance with his own wish expressed years before his death, his remains were placed beside those of his old botanical friend, Hobson, in the grave-yard of St. George's Church. The surviving botanists of the district assembled on the occasion of the funeral, to do honour to their late distinguished associate. Four of them—Thomas Heywood, George

Hulme, John Howarth, and James Percival, jun.—bore him to the grave. Of those who followed, including John Horsefield, Richard Buxton, Thomas Townley, Jethro Tinker—all men of local renown—many were advanced in life. "Three," says an account written at the time, "were sexagenarians; three others approached the period assigned of old for the duration of human life (threescore years and ten) and one who had seen about eighty winters, was still vigorous both in body and mind."

Nearly all the Lancashire botanists who flourished in the first half of this century lived to be very old men. The pursuits to which they devoted the leisure of their lives were wonderfully promotive of health and longevity.

The career of old Crowther was commemorated in verse by his son Richard. The lines were printed for private circulation, and, although possessing little artistic merit, they bring out one or two interesting traits of the man. They record how, intent upon nothing but filling his vasculum upon a long journey:—

> "He'd travel thirty miles on half a shilling,
> And homeward wending with the western ray,
> Would deem such toil and fare a well-spent day."

It is moreover recorded that, in addition to his love of plants, the old man took an interest in the collection of insects, minerals, and old coins, and was a student of the Lancashire literature of Tim Bobbin and other authors.

JOHN MELLOR, OF ROYTON.

THIS remarkable man, who, like many other naturalists in humble life, was originally a hand-loom weaver, died on the 5th October 1848, at the age of eighty-two. He was regarded by many, in his old age, with veneration as the father of Lancashire botany. Of his early career little is recorded. No doubt, amid the troubles with which the century was ushered in, he found less satisfaction in political agitation than in the practice of those homely virtues which best tend to promote human happiness. His great characteristic, even to the close of his long life, was a remarkably cheerful disposition, which, being united with a kindly nature and an insatiable thirst for knowledge, made his society much valued by men of kindred taste with himself. His earliest companions were John Dewhurst and George Caley, with whom he had many delightful excursions; and he became acquainted also with Edward Hobson, whose friend he remained until death parted them.

Mellor was not satisfied with the limits usually assigned by his fellow-botanists for their rambles —that is, the district of South-East Lancashire, North Cheshire, and the neighbouring portions of Yorkshire and Derbyshire—but he made frequent raids upon Scotland, and even penetrated into the Highlands. All such journeys he accomplished on foot. They were productive of rich stores of native plants—the envy of many who were not possessed of his hardihood or power of physical endurance—specimens of which he furnished on his return homewards through Glasgow and Edinburgh to Mr. Don and Dr. Hooker. Both these gentlemen well knew him and held him in high esteem. Having no independent income or other means of defraying the cost which his long journeys involved, he procured money always sufficient to reimburse him by sales of rare plants for cultivation. Six times he traversed the Highlands, and ascended Ben-Nevis, Ben-Lawers, Ben-Lomond, and the Breadalbane, Clova and other mountains, which are the homes of our rarest Alpine plants. He invariably preserved specimens for cultivation in his own garden at Royton.

During one of his journeys into Yorkshire, Mellor became acquainted with the gardener at Wentworth, who introduced him to the noble proprietor of that domain, Earl Fitzwilliam. The noble earl treated him with great respect, and purchased many rare plants from him. But his luck elsewhere was not so good. On one of his Scotch excursions, after visiting Edinburgh and Glasgow, on his return from a fatiguing journey in the Highlands, he was in possession of 13*l*. the produce of the sale of plants. An examination of the pocket in which he kept his money revealed the possibility of its soon wearing into holes, so, as a temporary expedient, he tied a knot upon it, enclosing three sovereigns, probably all the gold he had. The precaution thus taken was a happy one, for in the neighbourhood of Kilmarnock he was attacked one dark night by two villains, who robbed him of every penny that he possessed, except the three sovereigns, which escaped their observation. With these he was enabled to reach home. He ever afterwards had a horror of the Kilmarnock people, and in his journeys north gave that town a wide berth.

Mellor early abandoned weaving for a more congenial occupation. He spent at least the latter half of his life as a working gardener, and within a week of his death (he succumbed to an attack of British cholera) he might have been seen working in his own bit of ground at Royton.

So highly was Mellor esteemed by his old botanical associates, that a large number of them from all parts of Lancashire and from the adjacent county of Yorkshire, attended on the occasion of his funeral—which took place at Royton—to pay him a last tribute of respect. "It was pleasing," says a newspaper account, "to observe the true and heartfelt respect which these devotees of science paid to the memory of their venerable and truly estimable associate. Many of them had come a considerable distance on foot on a rainy day, and had sacrificed a day's wages, in order that they might accompany the remains of their old friend to the silent tomb.

We have often seen funerals conducted with more pomp, but seldom one with so much sincere regret and simple disinterested affection to the memory of the departed."

In his declining years this good old man, like too many of his class, had to struggle hard with poverty. But for the benevolence of Mr. John Roby, of Rochdale—the author of the 'Traditions of Lancashire,'—nothing would have saved him from ultimately seeking shelter in the workhouse.

RICHARD BUXTON, OF MANCHESTER.

RICHARD BUXTON, the author of the 'Botanical Guide,'* "a poor man" as he describes himself, "who had the greatest difficulty in procuring the necessaries of life in a worn-out trade—that of a child's leather shoemaker—and in delivering a few newspapers on a Saturday," was nevertheless an excellent botanist, and was acknowledged by at least one eminent scientific man (Mr. E. W. Binney, F.R.S.) to be the most profound thinker of his class. This man was born in 1796 at Sedgeley, near Manchester. His father was a farmer, who, whilst Richard was still an infant, fell into reduced circumstances, and afterwards followed the occupation of an ordinary labourer, living in an obscure part of Manchester.

The educational facilities open to a child under circumstances like these, may be supposed to have been extremely slight. Buxton thus describes them: "When quite a child I went for three months to a dame's school near my father's house, but I have no recollection of having been there taught anything. At the age of eight I went to a Sunday-

* 'A Botanical Guide to the Flowering Plants, Ferns, Mosses, and Algæ found indigenous within Sixteen Miles of Manchester.' —London: Longman and Co., 1849.

school in Gun Street, Ancoats"—the district of Manchester in which his parents lived—"I remained here upwards of two years, but did not attend regularly; and the furthest extent of my learning was being able to distinguish the letters of the alphabet, and to spell words of one syllable." His parents being unable to pay for his schooling, and his physical constitution being delicate, Buxton acquired a habit, while very young, of wandering through the fields and brick-yards near where he lived, to collect wild flowers. His favourites were the Germander speedwell (*Veronica chamædris*), the creeping tormentil (*Tormentilla reptans*), and the common chickweed, with its starry blossoms—plants common enough everywhere, but affording gratification to his infant mind, the recollection of which clung to him through life.

At the age of twelve the necessity was forced upon him of turning out to earn a little money for the household (he was the second of a family of seven children) and his father, with, apparently, little forethought, placed him with a man who followed the declining trade of a "bat-maker." For this man he continued to work a year and a half. He held his next situation for a period of several years. During all this time his education was utterly neglected; all that he had learned at the Sunday-school was forgotten; and at sixteen years of age he found himself utterly unable to read. His ignorance was a source of bitter regret to him, especially as he saw the rest of the family in possession of the coveted acquirement—that of reading—and, with commendable determination, he set about teaching himself. He got a common spelling-book, and, having mastered that, proceeded to read the New Testament; and as an aid to pronunciation, he procured Jones' 'Pronouncing Dictionary,' and went steadily through it from beginning to end. By this means was he not only enabled to read but also to understand the meaning of what he read, and to speak it correctly. He afterwards read historical books, and familiarised himself with the history not only of his own country, but also of Ancient Greece and Rome.

At the age of eighteen Buxton had the advantage of being in the service of a man—one James Heap—who had a certain regard for plants, but who cared to study more their medicinal uses than their scientific distinctions. Heap and his journeyman had long walks together, and the former brought home bundles of ground-ivy, centuary, mountain flax, and other humble but familiar herbs, of which he made decoctions, which he drank himself and "freely and gratuitously" distributed among his neighbours. In plant collection of this kind there is little to excite admiration; but it was, in Buxton's case, a preparation for something better. It could not fail to bring the young student into contact with species previously unknown to him. Plants were met with of which the names were unknown, and which were a serious puzzle, in their rambles, both to master and man. Culpepper's Herbal was resorted to, as a means of supplying the required information, but its rude and inaccurate descriptions of plants, and stories about their fancied virtues, soon caused Buxton to lose confidence in it. A systematically arranged Flora was what he desired; but he had heard of none, and the Linnæan classification was totally unknown to him. His first knowledge of that system was derived from Meyrick's Herbal, which had the advantage over Culpepper's of giving the Linnæan names and descriptions of the plants of which it treated. For years it continued to be Buxton's only handbook, but it, too, failed to satisfy him; and only by means of the Floras of Jenkinson and Robson, of the 'Botanical Arrangement' of Withering, and finally of the 'Introduction to Botany' of Sir J. E. Smith, was he, in the course of years, enabled to build up that reputation for scientific accuracy which he ultimately acquired.

His studies were long pursued alone. The men who had attained celebrity in their various spheres —Caley, Hobson, Crowther, and some others—were unknown to him. It excited Buxton's astonishment, in after years, that he should have remained so long unacquainted with those men, because they were making similar excursions to himself. It can only be attributed to the accident of their never having met; because it is inconceivable that men meeting on some wild moorland or mountain side, in the common pursuit of a study like botany, should fail from that moment to become intimate friends.

Another reason why Buxton failed to make the acquaintance of maturer botanists earlier than he did may be found in the fact that he did not, for a long time, identify himself with any of the societies of the district. How he formed an acquaintance with one shall be told in his own language: "One day in the month of June 1826, I went to Kersal Moor, and was quietly botanising on the back part of it near the brook, at a place now (1849) drained and cultivated, where grew a number of my favourite plants, when I happened to see a person engaged in the same pursuit. I made up to him, and asked if he was botanising. He replied: "Yes; I am doing a little in that way." I said that I had paid some attention to the subject myself; it was a study which had afforded me a good deal of pleasure. We walked about together on the moor, talking on botany and observing the different plants as we went along. .... I found him just such a man as I had long wished to meet—not a mere country herbalist, but an excellent scientific botanist. This was no other person than John Horsefield, hand-loom weaver, of Whitefield, the president of the Prestwich Botanical Society, and now the president of the general botanical meetings held at different places in the country lying between Clayton, Middleton, Newton Heath, Radcliffe, Eccles and Manchester—a profound botanist and well-read man, worthy to be the successor of Edward Hobson."

This meeting with Horsefield formed the commencement of a life-long intimacy between the two. It was more—it gave Buxton an introduction to some other enthusiastic botanists—all working men—amongst whom were James Percival, of Prestwich; Thomas Heywood, of Cheetham Hill; and John Shaw, of Eccles, on whose friendship he placed the highest value.

Association with these men, and more particularly with Horsefield, formed a new era in Buxton's life: it gave an impulse to his botanical studies, rendering them at the same time all the pleasanter; and best of all, intensified friendships which endured throughout life. It must have been an edifying spectacle, and one that we could desire more frequently, to see a number of hard-working unlettered weavers and shoemakers bound together as Buxton and his companions were by a common love of Nature, pursuing their studies with an ardour and intensity of enjoyment which nothing could allay.

It was not until the year 1833 that Buxton began to attend the general botanical meetings in Manchester. He found not only the men just mentioned but many other "pleasant and intelligent companions," amongst the members. At one of the meetings, held at Blackley, he became acquainted for the first time with James Crowther, "whose lively and cheerful manners," he says, "made such an impression on me that I ever afterwards enjoyed as much of his company as I could." Crowther, he adds, "was a man of good natural parts, and had a fair acquaintance with several branches of natural history; but his heart was far superior to his head. Hundreds of miles did we wander over the country agreeably together." About that time, also, Buxton found another valued friend in George Crozier. This was in the course of an excursion with Crowther and eight or nine other operative botanists, to Mere, Cheshire, in search of *Limosella aquatica* and *Elatine hexandra* which Crowther enjoyed the merit of having discovered some time previously. "All the plants," says Buxton, "that he had described Crowther showed us; and he appeared to enjoy quite as much pleasure in showing them as we did in having them shown to us."

This kindliness and generosity of Crowther is noteworthy. The incident shows how strong was his faith in the disposition of his fellow-botanists not to eradicate rare plants. A party of a dozen working men, all ardent botanists, could in those days walk thirty miles to observe rare plants in their native habitats, and gratified with the sight could return home and leave them still growing. If each one desired to possess a specimen, at all events care must have been taken—indeed was taken—not to imperil the growth and multiplication of the species, for they exist in the same spot to this day. Crowther was thoroughly unselfish. Many men having made the discovery which he did would have kept the secret to themselves.

Another friend whom Buxton found through his attendance at the botanical meetings was John Mellor, of Royton. He first met him at a meeting at Blackley, and was greatly pleased with the rare plants which he (Mellor) exhibited as the product of a recent journey to the Highlands.

Excursions into Derbyshire through the pleasant valleys of the Derwent, the Dove, and the Wye, gave Buxton an acquaintance with many plants growing in the limestone districts which he had never before met with. Other excursions into Wales and the district of Craven (Yorkshire)—rich in ferns—were fruitful of pleasure and interest.

For many years Buxton's botanical researches were carried on with unflagging zeal and great interest and pleasure to himself. He attended, in the year 1839, the meetings of the Botanical Class which was formed at the Manchester Mechanics' Institution upon the dissolution of the Banksian Society, and there found himself associated with his old friends Crowther, Crozier, and many other persons previously strangers to him. Amongst the latter was a gentleman (whose name he does not record) who had just begun to study botany, and appeared very anxious in its pursuit. "For him and at his expense," says Buxton, "I made many journeys into North Wales, Derbyshire, Craven, and other places." He continued to attend the meetings in question until, "in consequence of some dispute"—of the nature of which we are not informed—the class was unfortunately broken up. Still he made new friends. In the course of his journeys undertaken for the gentleman just alluded to, his attention was directed to the difficult genus *Carex*, of which he had undertaken to furnish his friend with specimens illustrative of the species growing about Manchester. In pursuance of this engagement he made tracks in all directions, and succeeded in adding fourteen species to those already known as belonging to the Manchester flora. In one of these excursions he was introduced to "that excellent botanist John Martin, of Tyldesley, weaver" (such was his estimate of Martin's character), who had paid particular attention to the *Carices*, and who freely communicated to him all the valuable information he possessed.

The next botanist in humble life with whom Buxton was fortunate enough to meet, was John Nowell, of Todmorden, a "twister-in" in the employment of Messrs. Fielden Brothers. "This man," he says, "in my humble opinion stands foremost among the working men with whom I am acquainted as a muscologist; and his extensive and accurate knowledge, joined to an excellent disposition, has always made his company a source of pleasure to me." Nowell and Buxton remained firm friends; and it is deserving of record that when, overtaken by old age, an offer was generously made to them by the late Sir William Jackson Hooker that they should spend the remainder of their lives at Kew, both refused it, because they could not tear themselves from the associations of Lancashire. This, in the case of Buxton—looked at from a worldly point of view—is remarkable, because he was at the time reduced almost to poverty by the extinction of his trade. Another remarkable fact is that, with all his love for the country, he continued to reside from boyhood in one of the poorest and most densely populated districts of the city of Manchester. He was living at the age of sixty-five in a little cottage in Gun Street, Ancoats, a few yards only from that in Bond Street to which, when he (Richard) was a boy in pinafores, his father removed from Prestwich. Having never married, he lodged with a married sister for a number of years before his death.

Buxton was a man of amiable and unassuming disposition. In the preface to his 'Botanical Guide' he disclaims any ambition to appear before the public; and says that, had he been allowed his own way, his book would never have seen the light. "But," he adds, "at the request of friends who in my opinion think more favourably of my attainments than they in justice probably ought to do, and who state that my book and history may possibly do some little good to my fellow working men, by showing them that the poor can enjoy the pleasures of studying science equally with the rich, I am induced to come out from my obscurity, and not only give to the public what knowledge I have acquired, but also state under what disadvantages and difficulties I have pursued my favourite science, botany." Of he friends above alluded to probably the foremost was Mr. Binney, who held Buxton in high esteem and gave him much kindly help.

For a number of years it was Mr. Binney's custom to entertain his humble friend, and all other naturalists in humble life living within a considerable radius of Manchester, to a good Christmas dinner. These reunions, which were discontinued on Buxton's death, were of the most jovial character. They were seasons of genuine happiness, every one having a sense of freedom from restraint, and each taking a delight in imparting scientific knowledge, in which all were interested. To none were the meetings a greater pleasure than to Mr. Binney himself, who has still recollections of the pleasant hours which he spent with these men. They never quarrelled, and in his opinion they had fewer jealousies than are to be found among scientific men of greater wealth and eminence.

The 'Botanical Guide' was the second Flora of the Manchester district which was published. It came out in the year 1849. The 'Flora Mancuniensis' preceded it by some years, but was a somewhat crude catalogue. It was edited by Dr. J. B. Wood, but was wholly dictated, or nearly so, by Buxton, Crowther, and other working men. The names of Buxton, Crowther, Crozier, and Horsefield, constantly appear in it as authorities for the recorded localities of rare plants. It is certain that but for these men botanists of higher standing and more liberal education would not have had so intimate a knowledge as they possessed of the Manchester flora. Dr. Wood, in his preface, bears testimony to the value of their researches, and says: "Perhaps in no part of the world is there a greater number of this class of naturalists to be found—in the same extent of country—than in Lancashire." To these men he dedicated the work. He expressed his obligation in particular to Richard Buxton, "whose retiring disposition and the obscure sphere of life in which he has moved" had "prevented his attaining that rank and estimation among the naturalists of the age to which he was so eminently entitled, by his profound acquaintance with Nature, acute powers of perception, and untiring industry."

Mr. T. Glazebrooke Rylands, F.L.S., in a letter to the author, refers to a botanical ramble which he once had with Buxton and Dr. Wood to Cotterill Clough. "Buxton," says Mr. Rylands, "was a character, and caused us much amusement. We could not get him to the train for looking at the good things he had gathered. As a last resource Dr. Wood threatened to throw his vasculum with its contents into the Bollin if he opened it again before we were in the train." This was characteristic of Buxton. The incident shows what an affection the old man cherished for the floral acquaintances of his youth. Upon him age had wrought its changes—physically at least—but they were the same, and capable of yielding the same enjoyment as ever.

Buxton was a man who never wrapped himself up in a mere selfish enjoyment of Nature. What he knew he was ever ready to communicate; and the pleasures he experienced in his botanical rambles were never so great as when there were others to partake of them. In his 'Botanical Guide' he appeals to landowners to allow the pent-up dwellers in cities to enjoy the beauties of creation, in fields and woods, without molestation.* "The fields and woods, although the rich man's heritage, may still be the poor man's flower-garden." He calls upon the sons of toil, living in the crowded streets and alleys of Manchester, to go forth with their wives and children and study Nature in her own retreats. "Best of all," he adds, "would I recommend them to avoid doing, or permitting to be done, any injury to the property of the owners of the land near to footpaths, or of parties who may kindly permit them to roam over their fields or wander through their woods. Let them know that working men can not only admire the beauties of Nature, but also thoroughly respect the rights of property."

Buxton retained, almost to the close of life, his love of long country rambles. At the age of sixty-two, according to his own testimony—although his constitution was by no means robust—he could walk

* There are few places where the humblest botanist will not meet at least with civility; but the author of this book has good reasons for advising Manchester botanists to avoid the domain of Wythenshawe.

thirty miles a day, and enjoy the beauties of Nature (as, indeed, may be conceived from the incident related by Mr. Rylands) with as much zest as he had ever done in his life. "True," he says, "the pursuit of botany has not yielded me much money; but, what in my opinion is far better, it has preserved my health, if not my life, and afforded me a fair share of happiness." This is the testimony of a working man addresed to working men, and is, on that account, of great value.

JOHN MARTIN, OF TYLDESLEY.

John Martin, a hand-loom weaver, whom Sir William Jackson Hooker, in one of his works, commends as an "accurate botanist," was, in comparison with the generality of the people amongst whom he moved, a learned man; and this fact, combined with his sterling honesty, gave him a local position and influence which few, perhaps none, of his contemporaries could hope to attain. In a brief autobiography, which he wrote some years before his death, he records that he was the son of a poor shopkeeper, who also kept a few cows, at Tyldesley; that he attended school from infancy until he was ten years of age; that then he was "put to the loom" for two years, upon which he had a further year's schooling; and that, beyond what he acquired in those his early days, he had the advantage of no systematic education. Having deep religious feelings, at the early age of fifteen, his great anxiety was to live secluded from the world, a "quiet and peaceable life in all godliness and honesty." His religious zeal, however, needed direction. Failing that, he lapsed into semi-scepticism. He made the precepts of the Bible his early study, and finding in the New Testament "passages inviting to voluntary poverty," he came to regard worldly riches with contempt. To lay up treasure upon earth was, in his view, a serving of God and mammon, and, therefore, the height of impiety. But he acknowledged that, whatever errors there might have been in his early religious impressions, those impressions had the good effect of keeping him from falling into dissipated habits, at a time when there was at least some danger of his doing so.

About the year 1800 Martin was obliged to leave home, in consequence of his father's failure in business; and two years afterwards he was married to a young woman of poor though respectable parents. "We were," says Martin, in the autobiography just referred to, "both of one age, to a few days; and our united ages were about thirty-seven years. About five years after marriage we had to provide for three children—all the family I ever had—and it was busy work to make all things comfortable."

The period referred to was succeeded by that direful period of distress and famine which was occasioned by the state of things upon the continent. The year 1812, and a great part also of 1813, witnessed intense suffering and privation on the part of the toiling masses in Lancashire. No foresight, no prudence, no care could avert the calamity. Food of all kinds was at famine prices. "Then," says Martin, "for the first time in my life, did I feel the miseries of extreme poverty. A certain writer remarks on such a case: 'Numerous miseries arose in view, and contempt, with pointing finger, foremost in the hideous procession.' I had been a regular contributor to the chapel which I attended, besides making occasional voluntary gifts; and I can truly say I gave beyond my ability, in the best state of my circumstances. Now, I was quite unable to pay my quarterly quota. The contempt consequent on my poverty I can never forget. Besides, the minister whom I attended, a moderate Calvinist, had an underhand and crafty scheme for getting an augmentation of his salary; and it was then I determined that, if I had any religion at all, it should be a religion without a priest, and that my chief object should be a study of Nature. In due time I withdrew."

Martin, at a very early period of his life, took delight in the cultivation of flowers; but it was not until the year 1816, or thereabouts, that he applied his mind with energy to the study of scientific botany. He was then over thirty years of age. Having access to good botanical works through the medium of a relation, who was a member of a local botanical society, and deriving some assistance also from "an affable old Scotchman, who delighted in botany," he made rapid progress. Having boasted to certain members of the society that he would outstrip them all, he deemed it prudent to delay his formal entry amongst them; but, nevertheless, attended their meetings occasionally as a looker-on. When he was admitted to membership, after the lapse of some months, he applied his mind to the study more energetically than ever, and for six years or upwards he read and studied nothing but botany—native plants, and those especially growing in his own neighbourhood, having his primary regard. The death of his wife, in 1825, was a serious blow. It almost led him to abandon the science altogether, but "certain other circumstances" forced him on again, and he pursued it with the same avidity as before.

Martin seems to have had only a slight acquaintance with the operative botanists then living about Manchester—at all events, he makes no mention of them in his autobiography—and amongst those living in his own neighbourhood there was a sad declension as years rolled on. In 1833 there was hardly a man in his acquaintance "who could look at a native plant," some of the older botanists having died, and others having abandoned the systematic study of native botany for the mere cultivation of flowers. These circumstances had little effect, however, upon Martin's ardour, for he continued his studies during many years, corresponded frequently with Sir W. J. Hooker and others, and undertook well-rewarded excursions in search of rare plants.

Into his domestic relations Martin gives us a little insight; and he goes out of his way even to express some odd views which he entertained on the subject of matrimony. He says: "Although I have never resided out of Tyldesley, during my passage hitherto through life, yet I have experiencend both pain and pleasure, sometimes to a great degree. These variations, both ways, have been caused chiefly through women! I do not complain. Take one thing with another, I have nothing to complain of." Again he says: "I have noticed the death of my wife, but have given no hint that I, after the lapse of twelve months, married another woman. To give a reason for this proceeding would, perhaps, be difficult, as marriage, I willingly admit, has no foundation in Nature; but sometimes things unreasonable become reasonable from their good results. This is Jesuitical, still it is true." His second wife died in January 1841.

Martin's services were much sought after, on account of his general good scholarship, by the local clubs and benefit societies. He was secretary of a men's sick-club for nearly half a century. In April 1807 he was chosen secretary of a female friendly society, and held office until its dissolution in June 1833; and in September 1820 he was appointed secretary to another female friendly society, which office he retained upwards of forty years. He never received remuneration at all adequate to the vast amount of care and trouble which he had to observe and endure; besides, he had to endure "all sorts of tempers, and many unkind and unjust reproofs." But he was of opinion—such was his tender regard for the gentler sex—that "Whatever faults or foibles there may be in the female character, they ought to be pitied and mourned over."

Martin was an amusing and graphic correspondent. The following letter is a specimen of his writing. It is well spelled throughout, written in a neat close hand, more like that of an advanced school-boy than of a man who had reached his sixty-second year. It was addressed to a friend, after a visit which the writer of it had paid to Manchester, where a natural history exhibition was being held:—

"Tyldesley, Feb. 9, 1845.

"DEAR SIR,—I now fulfil my promise by sending you some account of the impressions on my mind when I visited Manchester on the 6th of last month; but previously permit me to introduce some remarks on another subject, which I think is nearly allied to that which I mean to descant upon.

"Probably you are already aware that about sixty or seventy years since there was much controversy, among the *literati*, chiefly on the Continent, whether the savage or civilised state of man was most conducive to his well-being—that is, to his happiness. One writer, the Abbé Castrés, is pointedly and unreservedly explicit on the subject, maintaining that the savage state in the best: 'Knowledge,' he says, 'enervates our courage, multiplies our wants, concentrates man within himself, and insulates him from the public good; it narrows hearts and enlarges consciences, perverts morals and debases nature, and renders the efforts of virtue more painful. In short, the people degenerate as they become enlightened.' Then the Abbé praises the Caliph Omar for having set fire to the Alexandrian Library. But that the refined and accomplished Rousseau should have had the same sentiments to me is astonishing, Yes, even he, who during life had such an exquisite taste for the sublime and beautiful that, when dying, he requested his attendants to convey him to the window that he might have a view once more of his garden, and behold the splendour of the setting sun. Volney, on the contrary, in his 'Laws of Nature,' asserts that the man who is acquainted with the causes and effects of things provides, in a very extensive manner, for his preservation and the development of his faculties. Knowledge is to him, as it were, light acting on its appropriate organ, making him discern all the objects which surround him, and in the midst of which he moves with precision and clearness." Our countryman, Goldsmith —(I call him our countryman, although he was born in Ireland), maintains that knowledge and refinement are beneficial to man, and ludicrously alludes to the burning of witches, and other barbarous practices of our forefathers. Now, Sir, what must we say to these things? If the savage state is the best for the happiness and well-being of man, the sooner we retrograde the better. If the arts and sciences be pernicious, it is a pity we ever had them. Volney sums up in an admirable manner: "Preserve thyself, instruct thyself, moderate thyself; live for thy fellow-creatures, in order that they may live for thee."

"Assuming, then, the civilised state of man, with all its evils, is to be preferred before the savage state, let us contemplate Manchester, and its scientific institutions, as far as they impressed my mind on visiting them.

"From information, obtained from public and private sources, I had for several years been aware that there existed in Manchester a society for the study of geology, and that the members pursued the science with avidity; some of the members, if not many, being men of education, wealth and station. As I had myself paid some attention to the internal state of the earth (not scientifically), it made me a little anxious to have a view of their accumulated labours. On entering the museum, or depository, and seeing the vast number of specimens all arranged, I believe in the best scientific manner, I had a momentary regret that I could not understand the arrangement. The society has, I have no doubt, added much honour to Manchester; nor will the town be insensible to such honour. This, I think, is evident from the great respect paid to Dr. Dalton at the funeral of that eminently scientific man. My ideas respecting the Mechanics' Institution [exhibition] are a little confused. The objects being so numerous and interesting, I was rather bewildered: there was no time to examine anything in detail. When I saw the statues and paintings, my thoughts flew, with the quickness of lightning, to ancient Rome and Athens. I asked myself—How soon does a similar destiny await Manchester? But Manchester was not yet arrived at her full splendour. The fate of many former celebrated cities is well known. Many ancient cities which in former times were all magnificence, are now masses of ruins. Look at Persepolis, Palmyra, Nineveh, and many others. Surely there was a cause for all this ruin. I wish to believe that the knowledge of refined Europe has a tendency to mend the heart; but all my efforts cannot make me to believe it yet. Could savages do worse than our Government has done to the aborigines of Van Diemen's Land? The aboriginal inhabitants are all destroyed.—(See 'Chambers's Information for the People,' vol. i, page 53; printed in 1842). The writer says—'Safely, indeed, may we prophesy that in New Zealand, ere many years pass away, the natives will have disappeared before the European colonists. Not many months ago the last native thus disappeared from Van Diemen's land. So will it be, ere long, with New Holland, large as that continent is.' This is horrible.

"There was one thing worth going to Manchester for on the 6th January: what I saw at the Natural History Institution had a tendency to mend my heart. That was worth going for. To see all those animals in solemn silence which once were full of animation, each enjoying itself in its own native wood or place, to me was a good moral lesson. To see the Egyptian and Peruvian mummies was also instructing. From them I learned once more to look at the vanities of this life. I really wish I had not seen the remains of Miss Beswick. I cannot believe that she ever contemplated that her remains should be exposed to public gaze in the manner they are. If she did, I hesitate not to say she had a morbid imagination, and no attention ought to have been paid to her desires.

"Sir, I intended to have written much more, but am not well in health; besides anything that I have written, or could have written, would have been of little use, even if excellent—which it is not, and could not be. Neither for love nor money can I consent to have anything published that I write, either with or without my name.

When you have read this for your amusement, you can serve it as the Caliph Omar did the Alexandrian Library.

"From your humble servant,

"J. M. SAVARIUS.*

* "I assume this name from a maternal ancestor of the name of Savary."

The personal habits of John Martin were simple and inexpensive, even for a man in his humble position. He indulged in no sort of luxury. He was a strict vegetarian—at first, perhaps, of necessity, but ultimately from conviction—and, during the later years of his life, he not only abstained from animal food, but also entertained a strong dislike to all intoxicating liquors, which he never permitted himself to taste. "In stature," writes one who knew him, "he was of the middle size. Having lost the use of one eye, he wore a black shade, which in some measure assisted to show a fine massive forehead, indicating high intellectual capacity. His features were very fine, showing great strength of will, united to a most kindly and generous disposition—two characteristics of the man which can be amply confirmed by his old friends and fellow-botanists in humble life."

Martin died on the 13th August, 1855, in the seventy-second year of his age, having for some months previously been in a declining state of health, owing to an attack of paralysis. He was reduced to poverty, and it is said would have been forced to find shelter in the poor-house had not some gentlemen who were acquainted with his career given him some monetary help.

GEORGE CROZIER AND THOMAS TOWNLEY, OF MANCHESTER.

George Crozier, who possessed a high reputation as a botanist and entomologist, and who had also an extensive acquaintance with ornithology and other branches of natural history, was not, in the strict sense of the term, a working man, although some of the operative botanists of the district were his most intimate friends. During a residence of sixteen years in Manchester (where he died in the month of April, 1847) he followed the business of a saddler. His life was comparatively uneventful; but he had an amiable disposition, was unaffectedly cheerful, and at all times courteous—characteristics which gave him a place in the regard and affection of all with whom he came in contact. He was the life and soul of a large circle of friends, all devoted to natural history, and a companion, during many years, in their botanical pursuits, of Hobson, Dewhurst, Crowther and others, who were cut off before him. Buxton and Tinker were amongst the chief of his later associates. Thomas Townley, another able botanist, however, with whom Crozier had been almost through life on terms of fraternal intimacy, was his first instructor in the science of which he was so fond. Crozier — who was a native of

Eccleston, in the Fylde district of Lancashire, carried on business for some time at Blackburn; and it was during his residence there with Townley, that he commenced his botanical studies; but he removed to Bolton, and in a few months, again, to Warrington—in which latter town he added the study of entomology to his other pastimes—and finally settled in Manchester in or about the year 1831. During his residence in Warrington he numbered among his scientific friends Mr. Wilkinson, a tradesman of Bury, who is remembered as an accomplished entomologist. This gentleman was a passenger on board the *Rothsay Castle* steam-ship when she foundered in the Menai Straits, and was among the drowned—an event which was a source of great grief to Crozier. The friendship of Crozier and Townley was so close that when the former removed to Manchester, Townley was induced to follow him, and at the time of Crozier's death the two lived in close proximity to each other in Peel Street, Hulme.

Mr. Leo H. Grindon, the author of 'The Manchester Flora'—the latest and most complete Flora of the Manchester district—says of Crozier, in his 'Walks and Wild Flowers,' that few men have done more in their circle than he to foster and spread the love of Nature and of natural science. He was a prominent member of the Manchester Banksian Society, and of the natural history class which existed at the Mechanics' Institution for some time after that society was given up. The meetings, especially of the latter, were never complete without him; with him they were always a source of delight to the persons assembled. They were honoured, sometimes, with the presidency of Mr. J. Aspinall Turner (for some years M.P. for Manchester), always a warm and liberal patron of natural history; and by the presence of other gentlemen of position. The course of procedure was this: "After coffee had been served, short essays were read on interesting and popular subjects connected with the pursuit of the class; and from nine o'clock until half-past ten or so, the company promenaded, examining the curiosities in the glass cases that covered the walls, and enjoying the social pleasure which grows so largely out of soirées based upon a definite and intelligent idea, and where there is plenty to feast the eye. No one entered more thoroughly into the spirit of these gatherings than George Crozier. They were his festivals and harvest-homes, prepared for long beforehand, and looked back upon as isles of light and verdure in his wake. His love of social gatherings, and his skill as a practical naturalist, were equalled by his sagacity and shrewdness. 'There!' said he once, on the conclusion of the reading of a paper, 'that is what we want; that wasn't learnt out of a book.' His courtesy and generosity rose to the same level. Every Tuesday evening, when the members of the class assembled to compare their notes and discoveries of the past week, there

Crozier, busy as usual with his birds, and only too glad to chat with his young disciples, withholding nothing he could tell that would interest and amuse, and—what was far more valuable—inspiring them with his own enthusiasm. This kind, warm-hearted, cheerful old man it was who, taking the young naturalists by the hand, first showed many of them the way to Baguley and to Carrington, to Greenfield and to Rostherne, pointing out the rarities which his large experience knew so cleverly how to find, and communicating his various knowledge with the unselfishness of one in a thousand."

The extensive and accurate knowledge which Crozier possessed of plants rendered him expert in finding rare insects, being aware what species the latter fed upon. After a ramble with some entomological companions, less skilled than himself in botany, he would at times, while resting, institute a comparison of their respective "finds," and, his eyes sparkling with humour, would declare that his was "Benjamin's mess." "He showed, in the highest degree," continues Mr. Grindon, who wrote from a personal knowledge of the man, "how happy a man can make himself by the study of natural history, however humble his station in life, and however confining his employment. For Crozier, like Horsefield, Hobson, Crowther, and the rest of the Lancashire botanists, got his living by manual labour. He was a master saddler, and kept a shop on Shude Hill, the last place in the world one would look to for the abode of a naturalist, yet made, by his intelligent pastimes, one of the most contented in Manchester." A fellow Banksian described him as "one of those plain, plodding, practical naturalists, whose knowledge the field and forest, the uplands and the watery cloughs, had far more contributed to give than the lore of books. . . . The quiet, unromantic study of books would never have made either him or them what they were. Active adventure, real life within the whole domain of Nature, was their condition of enjoyment; and consequently the secluded footpaths, the fine old green and lonely lanes, the umbrageous bosky dell, with its clear babbling brook, and rich with plants, insects, and minerals, were their haunts." Who cannot echo Mr. Grindon's exclamation, after quoting the letter which contained these words:—"Would that our working men would, but a tithe of them, go and do likewise! Nature is just the same as when George Crozier loved it; the saddlers of to-day—the joiners, the weavers, and all the rest of them—are every bit as well able to enjoy and profit by it. In many respects their opportunities are superior. Go out, men, into Nature, and strive to learn and understand something about it; and if it be but a single fact in the course of the whole walk, you will have made a beginning, and have found how easy it is to become rich in pleasant knowledge; and, as we have said before, with nothing to pay except now and then a shilling or two for a little book."

"Never," says the same author, elsewhere, "was there a better example of the scientific man in humble life, or of the practical kind-heartedness and generosity that spring from simple, God-fearing virtue."

It was a peculiarity of Crozier, for a long time, always to wear a white cravat, which gave him a clerical appearance, and led to his being mistaken, at times, for a Methodist preacher. On one occasion, when on a journey to Cotterill Clough, with his son Robert, then a boy, trudging beside him with a tin vasculum, about the shape and size of a family Bible, some rustics, after directing him the way to Ringway Chapel, innocently inquired if he was "goin' a preachin'." On that point he did not enlighten them. Though he never assumed to be a preacher, he was, nevertheless, a willing listener in Nature's temple, always ready and apt to learn the great truths therein taught, and to lead others to an enjoyment of the same. His excursions to Cotterill were frequent, and were undertaken as much from a wish to enjoy the singing of innumerable birds, as from a love of the plants that grew there.

The following letter from Mr. William Wilson, addressed to Crozier, shows the relation in which the writer stood towards the operative botanists of his native county:—

"Paddington [Warrington], June 12, 1832.

"DEAR SIR,—I have just received an obliging summons from your friend, Mr. Tinker, to join him and you at Staleybridge on Friday morning, in order to spend that and the following day in botanical amusement. On such occasions I make it a rule never to disappoint civility by non-acceptance; and even were I less disposed, on other accounts, I should feel myself bound to accept the invitation, and shall accordingly determine to repair to Staleybridge at the appointed time, namely, Thursday night; or if prevented from that, to be there at the same time as yourself (Friday morning at eight o'clock). Probably I shall bring the gig with me; and, if I do so, it will be pleasanter to me, and I hope to both of us, to ride with you from Manchester, where I should stay on Thursday evening; and on my arrival I will give you a call, to know your plans and sentiments.

"My engagements are numerous just now, and but for the kind urgency of your friend, I should probably have been deterred from enjoying a pleasure which I have long hoped and wished for, without actually attaining it.

"We who make botany as much a leading business as an amusement, probably feel less gratification than you who make it an agreeable relaxation from the cares of life. At any rate, we are frequently obliged to attend to the less pleasing and more laborious occupations which belong to the study. According to circumstances, labour may become pleasure, or pleasure toil; they that eat most of the cake have not the highest relish; it is better to be temperate in all things.

"Yours, with much respect,

"W. WILSON.

"(Postscript.) — If unavoidably prevented from going, you shall hear again from me; but I don't intend to be stopped by anything."

Buxton and Townley have been mentioned as being, perhaps, the closest and most valued friends that Crozier had. The former, in the preface to his 'Botanical Guide,' describes, in a characteristic manner, his first meeting with Crozier. It was in the course of a journey to Mere mere, to see the rare plants which Crowther had discovered: "This ramble was of great interest to us all, but especially to me, as I then first made the acquaintance of George Crozier, now (1849), alas! no more, whose thorough love of Nature and kindly disposition greatly endeared him to me, and who continued, without ceasing, my good friend and faithful companion up to the time of his death." Had Crozier been spared, it was his intention to have assisted in the compilation of the work with which Buxton's name alone is associated, namely, the 'Botanical Guide.'

A memorandum relating to Crozier's last botanical ramble has been preserved. It was a ramble through some of the dales of Derbyshire. The memorandum was prepared before the journey was undertaken, by Buxton, whose superior knowledge of Derbyshire enabled his friend to make the most, botanically, of the journey. Localities of rare plants—some mosses among the rest—are pointed out, and precise directions given as to where *Verbascum nigrum*, *Carduus acanthodes*, and *C. eriophorus* in particular, were to be found; and the paper concludes with this reminder: "Bring Buxton two roots of *Polypodium calcareum* and *Verbascum nigrum*." This journey was undertaken in the summer of 1846. Crozier did not live to see the succeeding summer. An attack of bronchitis carried him off in the spring of 1847, in his fifty-fifth year.

Accompanying the memorandum just referred to, and some fragments of plants, there was found in Crozier's pocket-book, years after his death, a manuscript poem written by a man who subscribes himself "Black Jack"—no doubt an intimate friend of Crozier, but one, apparently, of strong radical proclivities. It is headed "Lines on Old Geordie," is addressed to "Friend Crozier of Saddling Notoriety," and seems to have been suggested by the death of George III. If the language is strong, let the character of the period during which that monarch bore sway, and the height to which political passion often rose, be borne in mind. The following is an extract only:—

"Here, then, we're equal—monarchs must
Their bones lay down to smoulder with the dust!
Few have occasion to lament their fate:
The little that they live's too long a date,
And, were it possible to bribe the grave,
They'd bury all the world themselves to save.
Few blessings on their subjects they bestow;
Their chiefest glory is their country's woe.
Mark the past scenes of Europe, and behold
Whole armies murder'd and whole nations sold.
But what are crimes in others are their fame:
War, famine, murder—all a royal game—
A royal sport—and 'tis too often found
The greatest murderer is the most renowned!
But mark their titles—'Mighty,' 'Gracious,' 'Just'—
Whose virtues are but gluttony and lust.
Their attributes by flatterers are adored,
And murder's mercy in a sovereign lord."

Crozier, though he had little sympathy with sentiments like these, was not the man to allow the conscientious views, either on political or religious subjects, of anyone who ranked as a naturalist, and whom he could claim as a friend, to cause an estrangement between them. He had his own views, both in religion and politics, but he knew how to respect the consciences of others. Townley, perhaps the most intimate friend he had, was a Roman Catholic. No matter, he was a naturalist too; and, as a naturalist, he had led him, through many pleasant paths, to the enjoyment of the pure and beautiful.

In a brief obituary notice of Crozier, which appeared in the *Manchester Guardian* of the 21st of

April, 1847 we read: "The last time the writer of this notice saw Crozier was at an annual meeting of the surviving Lancashire botanists, who dined together in Manchester in January; and he was much struck with the tall and upright patriarchal figure of the man — his intellectual head being covered with flowing white hair — his placid and intelligent features, and his cheerful, agreeable, and instructive conversation." The funeral of Crozier, like those of Crowther, Horsefield, and others, was attended by most of the surviving botanists within ten miles of Manchester.

Townley died in the month of September, 1857, having survived his friend ten years. He was a shoemaker. A more unselfish man, perhaps, never lived. During his residence in Blackburn, about the time when Crozier came under his tutorship, there existed a society of botanists at Mordden Water, of whom Townley was considered the leader, for he possessed a good library; and his great pleasure appeared to be in lending books to any or all who desired the use of them. There were few among his contemporaries who knew more of botany than Townley, yet, strange to say, he never formed a herbarium. Whilst studying plants, which he did with peculiar zest, in their native haunts, his great care was to store his mind with information concerning them, derived from the best sources. He addressed himself in an especial manner to the literature of natural history, and was of great assistance to Buxton, when that worthy botanist was engaged upon his 'Flora,' in preparing the notes with reference to particular species. As a tribute to the memory of his kind friend Crozier, also, he spent many days in arranging and cataloguing the extensive botanical collection which he (Crozier) left behind, and which Mr. Robert Crozier has—together with his late father's collection of moths and beetles—very carefully preserved. It was a labour of love on the part of Townley, whose catalogue and style of mounting and labelling are models of neatness and accuracy.

Townley numbered drawing among his accomplishments. He had considerable command of water colours—which he used chiefly for the portraiture of rare plants—and was, besides, an accurate sketcher. Under his tuition, when in Blackburn, Mr. Robert Crozier (to whom reference has already been made as an artist whose works have commanded a large share of attention) first developed a love for painting—a love which has descended to several members of his family, who are likely, ere long, to be heard of in the world of art. "It is pleasant to think," says Mr. Grindon, "that the beautiful pictures which now decorate so many walls had their impulse in the little palette of the old botanist."

An intense love of poetry, and especially of classic poetry, was another of Townley's characteristics. It is related, as an instance of the power of his memory, that he could recite long passages from any part of

Pope's translation of the 'Iliad,' without faltering. Such was his enthusiasm for Homer, that he used repeatedly to say, if his days were to be begun anew, he would learn Greek, if for no other purpose but to read that author in the original.

THE LESS-KNOWN NATURALISTS OF LANCASHIRE.

DURING the last half century there have lived in Lancashire a number of men closely associated with those already named—in some instances equally accomplished as botanists or entomologists—of whom it is impossible to give extended biographical notices. There was, in the first place, William Evans, of Tyldesley, a companion of George Caley and John Dewhurst, and correspondent of Dr. Hull and Dr. Withering. He was an enthusiastic botanist, and travelled many thousands of miles, it is said, in quest of plants. He was the soul of the early Tyldesley Botanical Society, and when he died, in 1828, his mantle may be said to have fallen upon his son, Joseph Evans ("Doctor Evans"), who still lives in the same neighbourhood, and is the president of the existing botanical society there. This worthy man has attained a good old age. His accomplishments as a botanist are acknowledged by Mr. Grindon in the 'Manchester Flora.'

WILLIAM WORSLEY, a weaver, was born in 1808, held a high position amongst the operative botanists of Middleton, and was an active member of the society which existed there in Caley's time. He was a man of delicate constitution, and suffered in health from having to live in a damp cellar, owing to extreme poverty and miserable food during a period of great distress. "This dark lot," we are told, "had one bright spot, for Worsley loved books, and was specially fond of botany. Remarkable from his youth for clearness and strength of mind, he spent in studying English grammar the time which other boys gave to play; and when he reached man's estate eagerly devoured whatever reading he could obtain. Among the works that fell into his hands was Culpepper's Herbal, the coloured plates of which stimulated his mind and led him to give attention to botany. His quiet and often lonely rambles in the fields and dells about Middleton gave him ample opportunities for exercising and improving his knowledge of plants." When most of the old botanists of Middleton were dead, the rising ones who during their walks had plucked a strange wild flower would bring it to Worsley to be described and named. Hence arose a botanical society, with Worsley as its centre, of which John Turner, a shoemaker (who corresponded with Caley during the residence of the latter in Australia), and Samuel Barlow, of Stake Hill, manager of a bleach works, were prominent members.

Samuel Carter, of Manchester, was another of these self-taught naturalists. He was a member of the Banksian Society, and of the natural history class which followed its dissolution. "Years before he learned anything of scientific natural history," says Mr. Grindon, "he had made himself thoroughly acquainted with one of the most curious and perplexing chapters of out-door zoology, namely, the phenomena of the life and growth of frogs. The result was that, when admitted as a Banksian, though only a lad among old men, in experience and genuine knowledge he was, in this respect at least, little behind them." Carter was an enthusiastic entomologist, and possessed one of the largest collections of British insects in Manchester. He re-arranged the entomological collection in the Manchester Museum in the year 1858, in which task he displayed an ability that could only be appreciated by those who knew its magnitude and difficulties.

Carter followed the trade of a cabinet maker, and was able to earn a sum of money which after his death yielded his widow a weekly pittance of eight shillings. She, however, in course of time, was induced to part with her interest in this money, by some fellow, on the faith of a promise of marriage, and she lost all. Ever afterwards she lived in a state of great indigence, latterly with a brother, who was in receipt of parish relief—and was utterly broken in spirit. She had one child, a daughter, who was married and resided abroad. Her brother died suddenly in the early part of this year (1873) and she survived him only a few months.

JETHRO TINKER, to whom allusion has been made in the preceding pages, died in the month of March, 1871. He was up to that time a sort of living link between the present and the past; and in his native town of Staleybridge, as well as in the surrounding towns of Ashton-under-Lyne, Oldham, Mossley, Saddleworth, Mottram, and Glossop, was held by naturalists in high esteem, as one of the early pioneers in the study of botany and entomology in that district. For a long time he worked alone and unaided, but with great ardour and perseverance, studying, whenever they came in his way, the works of Dr. Withering and other early English botanists. It was his good fortune, during a long lifetime, to make many botanical and entomological discoveries. He delighted to initiate others into a knowledge of the mysterious workings of Nature, and was ever ready to impart freely and liberally the knowledge which, by patient and unremitting toil, he had acquired for himself. He never made his knowledge or discoveries a means of gain. For many years prior to his decease, which was sudden and unexpected, he held the post of chairman of the Botanical and Entomological Section of the Staleybridge Naturalists' Club.

TWO LANCASHIRE BOTANISTS NOT IN HUMBLE LIFE: JOHN JUST AND WILLIAM WILSON.

ALTHOUGH the design of this work is to furnish a record of scientific men in humble life, the author feels that, when dealing with botany in Lancashire he may fairly be excused if he introduces some notice of those two ornaments to the county—both, alas! no more—Mr. John Just, of Bury; and Mr. William Wilson, of Warrington.

MR. JUST was the son of a Westmoreland farmer, and was a native of the village of Natland, about two miles from Kendal. He obtained the rudiments of an English education at the endowed school of his native village — then kept by Mr. James Ward, an artist of some repute—and as he grew up, being strong and robust, he was employed by his father in ploughing, harrowing, and other farm work. He early manifested an intense love and aptitude for learning, and was sent, when about fourteen years of age, to the Kendal Grammar School, where he commenced a classical education, which was afterwards continued for some years in the grammar school of Kirkby Lonsdale. He had there an excellent tutor in the Rev. John Dobson, whose classical assistant he ultimately became.

At this early period of his life he manifested a taste for antiquarian pursuits by investigating Roman remains in the neighbourhood, being stimulated, doubtless, by the fact of the existence of a Roman station at Natland, in a bend of the river Kent, called the Water-brook, and supposed to be the site of the ancient Concangium. The ramparts of a square fort were some years ago, and may be still, discernible at this spot; and various relics have from time to time been found. Once, when a pupil at Kirkby Lonsdale, he travelled on foot, during a winter's holiday, to Borrow Bridge, on the Lune—distant sixteen or seventeen miles—for the purpose of examining some fine Roman remains which existed there. Those only who know that mountain road can adequately estimate the arduous labour of the undertaking.

The late Mr. Harland, in a communication to the Manchester Literary and Philosophical Society (to which we are indebted for much of the present information), says that, in conection with his pursuit of natural history, Mr. Just used to relate many amusing anecdotes, unconsciously illustrating his own kindliness of nature, of his obtaining birds and small animals when a boy and keeping them for some time with a view of testing the effects produced on wild creatures by domestication and kind treatment. While assistant to Mr. Dobson, it was his habit to rise in the summer mornings at four o'clock, in order to pursue the study of botany, not from books but from nature. So thoroughly had he botanised the district that there was not a habitat of any plant at all rare within many miles of Kirkby Lonsdale with which he was not acquainted. He examined and named every plant himself, and when, in after years, he taught botany he made his pupils do this, urging that, unless they acquired the habit, instead of trusting to others, or to books or engravings, they would never make good botanists.

Mr. Just, after the lapse of a year or two spent in Mr. Dobson's employment, removed to Bury, and he was, in 1832, elected to the second mastership of the grammar school in that town—a post which he filled with credit during the remainder of his life.

His vacations were usually spent in pedestrian excursions, in which his love of Nature, his taste for agricultural avocations, his eager pursuit of botany, and his keen zest for antiquarian researches, had full scope. In the midsummer vacation of 1834, he made a tour of the Highlands. One evening he commenced, with a friend, at seven o'clock, the ascent of Ben Nevis, in order to be in time to enjoy the early morning prospect from the summit. The ascent was successfully made; and the friends reached their hotel at seven o'clock the next morning. There they found another party preparing to make the ascent, and so delighted was Mr. Just with what he had seen, and so ready was he to be useful as a guide to the strangers, that before nine o'clock—after a short rest of little more than two hours—he was again *en route* for the summit of the mountain. He accomplished the toilsome ascent a second time; and again reached the hotel between five and six o'clock in the afternoon, apparently "not more than comfortably tired" by the extraordinary physical exertions of the preceding twenty-four hours. He acknowledged afterwards, however, that in making the second descent he walked as if mechanically, and often seemed to himself to sleep whilst walking. This story is told as affording a fair estimate of the muscular strength and power of physical endurance of the man.

In the month of October, 1848, the lecture committee of the Royal Manchester Institution recommended that an honorary professorship of botany should be instituted, and that it should be offered to Mr. Just; and the council having unanimously confirmed the appointment, it was tendered to and accepted by Mr. Just, who expressed an earnest desire to discharge faithfully and diligently whatever duties might in consequence devolve upon him. Well did he redeem this pledge. In September, 1849, in his character of honorary professor, he delivered gratuitously a course of three evening lectures on botany at the Institution. In May and June, 1850, he delivered a more extended course of six afternoon lectures also on botany, and its various systems and classifications; and in the spring of 1851 he delivered a further course of six lectures on the various organs of plants. Ill-health compelled him, to his extreme regret, to abandon an engagement to deliver a course of lectures in 1852. On the eleventh of May in that year he wrote to a friend: "I have been an invalid upwards of nine weeks, and can scarcely walk; whether I shall get out again I consider very uncertain." His days were, indeed, numbered. He died in the autumn of that year, at the age of fifty-five.

Mr. Just's powers of discrimination, and of remembering the specific characters and properties of plants, were remarkable. In gathering any plant, even a moss or a lichen, which he had ever before examined—even in years long past—he would without any reference to books at once name it, and if his correctness were doubted he would point out its distinguishing marks and show wherein it differed from similar plants. The discovery of the habitat of any scarce plant was a source of intense delight to him; and nothing annoyed him more than the extirpation of rare species by collectors. He discovered a habitat of the lady's-slipper (*Cypripedium calceolus*) at Arncliffe, in Yorkshire; but unfortunately, on its becoming known, the plant was sought out by gardeners from the neighbourhood of Manchester, and rooted up. Many years after the discovery Mr. Just pointed out the locality to a friend, expatiated upon the delight which it had afforded him, and expressed strongly his regret that so rare and beautiful a plant should have been eradicated for the purposes of sale.

The friend who was with Mr. Just, when he found the *Cypripedium* (in, it was believed, the year 1835) thus describes the incident:—"It was on a fine morning in July. We were quartered at Arncliffe, and before breakfast had rambled into a scraggy limestone wood for the express purpose of looking for the lady's slipper, which we were told had been found there within the last ten years. Imagine the eagerness with which every nook both likely and unlikely, was searched. We had separated from each other a distance of perhaps twenty yards, when I was attracted by a joyous scream; at the same time I saw my friend's hat high whirling in the air, and, with a schoolboy's delight, he thrice shouted—'Eureka! Eureka!! Eureka!!!' And, sure enough, there was the envied prize—two plants, in beautiful bloom, and five small seedlings. We each brought away one of the blooming plants; and I am afraid the news of our success proved fatal, at length, to the seedlings."

The same botanical friend has recorded of Mr. Just that, "as a vegetable physiologist, he might be placed in the first rank, although he never extended his acquaintance with plants beyond those which were indigenous to Britain; but in every department of British botany he was thoroughly versed.

The plants of our own country afforded him a field ample enough to study the laws of the vegetable kingdom, which were the great objects of his research, and in which he made many interesting discoveries. To know the names of a great number of plants, or to have dried specimens pictorially laid down upon paper, was not what he termed botany. A precept of his which deserves to be remembered, and which he once repeated to me, when I remarked that I knew a particular plant from a certain resemblance that it bore to some other, was: 'If you wish to become a botanist, you must learn to distinguish plants by their differences, and not by their likenesses'—a piece of advice I have found useful on many occasions since. He was especially fond of cryptogamic botany, as exhibiting the wonders of creative wisdom displayed in these pigmies of vegetation; and many a new species was discovered by him before it made its appearance in any English work, or was considered as a British plant."

Mr. Just did not devote his attention exclusively to botany, much as he loved that science, but was a man of very varied acquirements. He studied farming in its scientific and philosophical aspects, applying to the study his extensive knowledge of the sciences of chemistry, geology, botany and even mathematics. He was the author of many learned archæological essays, and his contributions to the Literary and Philosophical Society of Manchester, enriched the pages of its 'Transactions.' A friend wrote of him: "If his circumstances, or the help of his friends, had given him [in youth] the means of pursuing those studies for which he had so great a taste, and accompanied with the advantages, in due course, of a university education, he would not unlikely have gone on in the same career of distinction with some of those painstaking scholars of the north who, like a Sedgewick or a Whewell, have gained for themselves the highest distinction by their contributions both to literature and science."

He relished country pastimes, and was from boyhood a keen and successful angler. It is said that many an experienced angler on the Lune and Kent would suspend his sport, in order to watch Mr. Just's more artistic and successful casts of the line. He looked forward to his annual fishing excursions with delight, not more for the sake of the sport than for the opportunities afforded him of visiting and studying the botany and antiquities of new localities.

Some scraps of Mr. Just's correspondence have been preserved, and are well worth quoting, as showing how closely wedded he was to Nature. Writing to a friend in April, 1834, he says: "You rally me on the gaiety of the life I am leading. It is far from being congenial to my feelings, and instead of adding to, subtracts very materially from my happiness. I often envy the days gone past, when alone in woods and wilds I took my solitary rambles, unnoticed by anyone, and conversing with

God and His wonderful works in nature alone. Such a course gives an elevation to the mind and a tranquillity to the conscience which not all the flattery of the great nor the approval of the wise in this world can produce. Often have my friends, when I lived in retirement in Kirkby Lonsdale, scolded me for the apathy and indifference which I manifested towards my own welfare in the world, and my advancement in society; but, in spite of all, I was then wise. The improvement of my own heart and mind was then the sole object of my ambition; and would it still were so! I would I had the chance of retiring into some sequestered vale with the prospect of a pittance for life just sufficient to keep me from the meanness of poverty without the power or the temptation of becoming rich; that every night I might lay my head upon my pillow with the satisfaction of knowing that I was becoming wiser and better every day."

"Nature," he again wrote, "is my garden; and who or what man or woman, or destroying spirit, can spoil it? I may have no pet place wherein to shelter my favourite flowers, but they are still as beautiful, in my eyes, wherever seen, wherever they may blow. Give me a cot in the country, and quiet; and I desire no more. The fields around will furnish me with delight, as I visit the flowers which none knows, or sees, but the industrious bees and myself."

The memory of this good man is tenderly

cherished by the few surviving friends with whom in his varied scientific works he was so long associated.

Mr. William Wilson, as a botanist, enjoyed a reputation hardly inferior to that of Sir William Jackson Hooker—whose intimate friend he was for many years—and others, whose names have been more prominently before the public. His *magnum opus*, the 'Bryologia Britannica,' published in 1855, is acknowledged to be the best work on British mosses that has ever been published, and at the time of his death, on the 3rd of April, 1872, a new and enlarged edition of it was in contemplation. As a bryologist, indeed, no man living was a better authority. The late Mr. Wilson's friend and correspondent, Professor Schimper, of Strasbourg, can be said only to have equalled him, and even that eminent man yielded the palm to our countryman in the matter of manipulation. It is said that when in Warrington, a few years ago, the Professor was astonished to find Mr. Wilson sketching with one hand and dissecting with the other, and he could not find words in English to express his feeling: "*Non equidem invideo; miror magis.*" The friendship of these men lasted to the end. The following letter, written by Professor Schimper after the siege of Strasbourg, has an almost melancholy interest :—

My dear Wilson,—I thank you much for your kind letter, in which you have kindly wished to express your sympathy with us in the sad circumstances which overwhelm us. You will understand how in the midst of the great misfortunes which have overtaken my country my mind is indisposed for scientific researches. During my absence, when the Prussians began to bombard the town, my friends had my library and my collections carried into the cellars. These now find themselves once more in their old place, but in great disorder, nor do I know when I shall be able to find time or sufficient quietness of mind to set them to rights. I have had to give up my study to the soldiers, and it will be out of the question to attempt even the smallest microscopical work. I had hoped to finish my great work on vegetable palæontology by the end of this year, and get back to my dear mosses at the beginning of the next; but now all that is interrupted. It is a great pity that you cannot get an assistant for your second edition of the 'Bryologia.' It will be a most unfortunate thing for science if the large quantity of material you have collected should be lost, but I hope that your health may be sufficiently restored to enable you to bring your second edition to a successful and happy termination.

"Your devoted

"W. T. Schimper."

Dr. Lindberg, of Helsingfors, another valued correspondent of Mr. Wilson, spoke of the 'Bryologia Britannica' as the most accurate book on mosses that he knew, and he wrote on the 10th January, 1871: "To-day I have to wish you a most happy new year, with health, large income, &c., and that you may publish your important new edition of the 'Bryologia' under this year."

Ten days later Dr. Lindberg again wrote: "I have you and your excellent 'Bryologia,' one of the most exact works in botany, to thank for so many instructions that I cannot be too thankful, against such an eminent *savant* as you are. And you yourself, too, have been so good and generous to me that I am in debt to you always. I wish only I had seen you in the very face; but the circumstances have hitherto been adverse. Perhaps they will change: I at least live in the hope."

Only a fortnight before Mr. Wilson died, the publishers of the 'Bryologia,' Messrs Longmans, wrote, saying that Dr. Hooker, of Kew (the son and successor of the distinguished Sir W. J. Hooker in the curatorship of the Royal Botanic Garden) had expressed an opinion that, as the 'Bryologia' was so frequently asked for, it would be desirable to reprint the first edition as it was, unless Mr. Wilson was likely to be able to prepare a new one in a very short time. The hopes, however, of so many friends, who looked forward to the new edition from Mr. Wilson's own hands—and for which that gentleman

L 2

had spent two years in gathering material—were cruelly disappointed.

With regard to Mr. Wilson's personal character, the *Warrington Examiner*, in a notice (which bears evidence of having been written by another highly valued scientific friend living in Warrington), says :—

"Of his character we speak with diffidence. In this, as in personal stature, he was (as he was often playfully called) Saul, the son of Kish, head and shoulders above all the people. Scrupulous conscientiousness was his great characteristic ; to live a humble Christian life, his most fervent desire ; to serve his generation in the work God had given him to do, the object of his life ; in all things to do rightly and justly, no matter at what cost or sacrifice, his aim and ambition. Fame or praise he never sought, for that he was too truly great ; he knew too well how little, in comparison to what remained undone, his best work was. But he was not cold or disdainful ; real appreciation was a comfort as well as a help to him. His estimate of himself was always a very low one, because the standard he placed before himself was so high; but though he disparaged himself, he had attained much in wisdom, in knowledge, in true piety, and real benevolence. He was brought up a Nonconformist, and from strong conviction held firmly to the principles and practices of the Congregationalists. In politics he was a Liberal, though he took no active part in political life. Always of delicate health, all through life he had to struggle with bodily ailment; he said hard things against himself, deploring what he called his irritability and impatience, making no allowance for the extenuating circumstances he would so willingly have allowed to anyone else. All who knew respected him; those who knew him best knew the lovingness and loveableness of his nature. He was dearly loved by children, and returned their affection. If they showed intelligence, he thought no pains thrown away in explaining to them his work and his scientific treasures: the few who had the *entrée* to his study will never forget it. Young men wishing to study natural science found no difficulty in approaching him; he was always willing to aid and encourage them, making himself a friend, taking an interest in all that belonged to them, rejoicing greatly in any success they might attain. He had a deep reverence for good women, and could never speak of his mother without deep feeling. On his death-bed he said that, when tempted to take the gloomiest views of human nature, his great comfort was to think how refined and educated women undertook the most menial and revolting services for the sick and dying, from the purest motives."

Early in February, 1872, Mr. Wilson caught a cold, which settled on his lungs and produced severe inflammation. This was subdued, but a distressing attack on the brain supervened. From this, too, he rallied; and during the last weeks of his life, his mind was unclouded, calm and peaceful;

but the strain on the body had been too great, and the patient sank from exhaustion. He died peacefully.

"How to speak of that death-bed," says the writer of the account already quoted, "we scarcely know. He never forgot his life work; a letter from some botanical friend never failed to rouse him to temporary energy. The only thing he wished to live for was to bring out a revised edition of his 'Bryologia.' Except for that, he longed and prayed to be at rest. His sufferings were very great; his patience was wonderful. No repining word escaped him, no expression that anyone but himself would have called impatient. Having one day said to a friend, 'Oh, do be quick!' when the paroxysm was past he asked forgiveness for his impatience, adding in his beautiful self-abnegation, 'It passes my comprehension why you should care to do anything for such an one as me.' His particularity never forsook him; he had asked one of his nurses for soda-water, and stopped to show her the proper way to manage that unruly liquid, saying: 'There is a right way in all things, and though I am but a learner in most things, I ought to know that one.' Another thing that showed the man was his anxiety not to give unnecessary trouble: it was no use to say you wished to do everything possible; he would reply, 'Because you are good I must the more carefully guard against taxing your goodness.' This is scarcely the place to speak of his true piety, but as that was the ruling power of his whole life, the cause, not the effect, of his botanical studies, we must not omit all mention of it. The few who saw the triumph of faith over doubt, who heard the earnest prayers, the deep contrition, the humble hope, of that dying man, can never forget that sick-room. He died as he had lived, a humble, trusting, not triumphant Christian. His triumph is on high. May our last end be like his."

His body rests in the ancient Nonconformist burial-ground at Hill Cliff. His only daughter placed upon his coffin, before it was lowered into the grave, a beautiful wreath of *Hypnum*, with a little bunch of violets and ferns—a tribute of respect and love rendered by Mr. George E. Hunt, of Bowdon, who was one of Mr. Wilson's youngest botanical correspondents, one in whom he took the deepest interest, but who survived him only one short year.

Of the late Mr. Wilson's early career, and the circumstances which led to his devoting a lifetime to the pursuit of botany, it may be expected that something should be said. He was a native of Warrington—the second son of Mr. Thomas Wilson, who carried on business in that town as a chemist—and gained the first rudiments of education at a dame school kept by a Mrs. Du Garney. This old dame is described as a character, in her way. She had been an actress, had married a French refugee, and prided herself on her elocution. Mr. Wilson acknowledged that he owed to her his first knowledge of, and love for, pure English pronunciation. His next tutor was a Nonconformist minister. Ultimately he was sent to the Prestbury Grammar School; and completed his education at the Dissenters' Academy in Leaf Square, Manchester, then under the management of Dr. Reynolds, who was both a gentleman and a scholar. As education went at the beginning of this century, he was well educated. Dissenters did not then send their sons to Cambridge or Oxford ; but they did educate soundly and thoroughly, so far as they went, in their own academies.

The legal profession was chosen by his father as that which Mr. Wilson should pursue; and, though it was averse to his tastes, yet when he found himself an articled pupil of Messrs. Barratt and Wilsons, of Manchester, he applied his mind so closely to the study of conveyancing as to bring on distressing headaches, and lay the foundation for a serious illness. He found it necessary to take protracted holidays and make long journeys, in the hope of regaining his health.

From a boy he had loved Nature and all her marvellous works, and now in his lonely rambles took up the study of botany as his special amusement, his younger brother taking up entomology. About the year 1824 or '25, Mr. Wilson's mother, who seems to have understood her son, offered to give him a sum of money, the interest of which should enable him to pursue his favourite journeyings in the very modest and economical way in which he had begun them. He hesitated to give up the law, and, as it seemed to him, throw away the money spent on his preparation. Increasing illness obliged him to yield, and the following letter from Sir James Edward Smith, then the chief botanist in this country, quite decided him to make botany, not his relaxation only, but his life's work.

"Norwich, May 10th, 1826.

"DEAR SIR,—Overwhelmed as I am with letters, queries, commissions—of which I have been labouring to get through what have accumulated during last summer's absence, amid bad health, and heavy family afflictions; when your parcel came I exclaimed—'More letters to answer! The more I write the more I receive!' I had not for some moments courage to open it. But when I did, how was my tone changed! Instead of idle questions, I found such an assemblage of varieties and novelties as have rarely met my eyes, accompanied with so much excellent intelligence and such kind offers as made me put everything else aside and resign myself to that pure pleasure which botany, 'in sickness and in health,' in trouble or prosperity, has so often afforded me. Be pleased, therefore, Sir, to accept in the first place my grateful thanks for your liberality and kindness to an entire stranger, but (I hope) to one who will not prove unworthy." Two closely-written pages of botanical lore follow, and the letter concludes thus: "Any important corrections or remarks cannot but be welcome from your pen."

So cordial and appreciative a letter from so high an authority was just what the modest young man needed.

Mr. Wilson was not without encouragement from other quarters. On the 20th of May, 1826, Professor Henslow wrote: "I have the greatest pleasure in accepting your offer of becoming my correspondent." Three other letters in the same year acknowledge the receipt of valuable specimens and information. On the 17th of April, 1827, Professor Henslow introduced to the young botanist "Dr. W. J. Hooker, Professor of Botany at Glasgow," who desired Mr. Wilson's valuable aid in a forthcoming number of the 'Flora Londinensis.' On the 16th of May Dr. Hooker wrote acknowledging prompt and valuable help, pointing out fresh fields of work, and specially recommending the study of mosses. In the same letter Mr. Wilson was invited to join the Professor's class in a five-days' botanical excursion among the Breadalbane Hills. This welcome invitation he accepted—could an enthusiastic young botanist have done otherwise?—and he prolonged his stay in the vicinity of Killin till the middle of September. This was the commencement of a life-long friendship and fellowship of work with the distinguished Professor, who, on the 18th of September, wrote to Mr. Wilson's mother assuring her of her son's improved health, saying that he had sailed from Glasgow to the Isle of Man, and adding: "He was so agreeable an inmate of our house that I assure you both myself and all my family are looking forward to his paying us some future visit with very great pleasure. There are few persons in whom I have felt so much interest upon so short an acquaintance."

After this Mr. Wilson spent two years in Ireland, and made some of his most important discoveries. Though Mr. Wilson published nothing in his own name until 1855, yet as early as 1829 Dr. Hooker writes that he looks to him for assistance in his English Flora : and in turning over the pages of that book one sees continually such remarks as—"I am glad to have Mr. Wilson's authority for this;" "I am supported in this view by Mr. Wilson," &c. &c. Dr. Hooker's letters contain constant, repeated, and various acknowledgments of assistance; of specimens, of information, and frequent praise of Mr. Wilson's steady, continued, and beautiful work. So the years rolled on, one great man after another seeking introductions till William Wilson's name became known all over the world. At the end of his life, as at the beginning, letters came overflowing with thanks for help, acknowledging liberal gifts of specimens, asking advice or confirmation. The names are changed, but the same appreciation of the man's genius remains.

In 1836 Mr. Wilson married his cousin, Mrs. Lane,

at St. Pancras Church, London. The modest income which had served for bachelor days was now needed for family claims; and his journeys became less frequent and more curtailed. Orford Mount (Warrington) was his residence for many years, and much of his best work was done there. On his mother's death he removed to the house at Paddington, about two miles from Warrington, in which he died.

His study there was a sight to think about. His order and neatness were very remarkable, but he had no expensive arrangements. The lesson of that plainly-furnished room was what great results might be obtained from very insignificant means—how the seeing eye found uses for the most trifling objects.

SAMUEL GIBSON, OF HEBDEN BRIDGE.

FEW naturalists in the humbler walks of life have enjoyed a higher reputation than did Samuel Gibson, who to his varied attainments as a botanist and ornithologist, added a very intimate acquaintance with the sciences of geology and mineralogy. It was, indeed, as a geologist that he was best known in the world of science. He was a close observer of Nature. He may, indeed, fairly be classed as a discoverer, for it is through his careful investigations in the Vale of Todmorden that geologists became aware of the existence of many fossils in the lower coal measures, of which previously they had been ignorant.

A brief memoir of Gibson, published shortly after his decease, which took place on the 21st of May, 1849, says: "He was an indifferent penman, and having to work hard at the laborious trade of a whitesmith for the support of a family of nine children, he had little leisure for writing. . . . . Although comparatively little known in the world, many valuable facts observed by him have enriched treatises on natural history"—in proof of which latter assertion it is only necessary to refer to Professor J. Phillips's 'Geology of Yorkshire.' The

Professor, in his introduction to the second part of his elaborate work, expresses his great obligation to Samuel Gibson, and also to Mr. Francis Looney, F.G.S., of Manchester, for their prompt attention in sending him "specimens of many fossils from new localities in the limestone shale of the Vale of Todmorden." In order to mark his appreciation of Gibson's labours, the Professor caused a species of Goniatite which he (Gibson) found to be named *Goniatites Gibsoni*. Figures of this and other species which Gibson had the merit of discovering appear in the course of the work.

The best testimony, however, to Gibson's attainments—the best memorial, we should rather say, of his high qualifications—is to be found in the geological department of the Natural History Museum (now the property of the Owens College) at Manchester. There, in careful preservation, may be seen the splendid collection of fossil shells from the lower coal measures, of which a description, written by Captain T. Brown, with beautifully engraved figures of the several species, appeared in the first volume of the 'Transactions' of the Manchester Geological Society. Some of them are unique; and a large number Gibson's own discoveries. Captain Brown thus refers to Gibson's labours, after expressing his obligation to him for the loan of his collection of new and interesting shells: "Situated in a country village, remote from men of science, and books, and destined to earn his bread by a laborious employment, this intelligent and excellent individual has by his personal energies, apprehension, and great industry, overcome all the difficulties which beset him, and has done more in the way of collecting new objects in almost every department of local natural history than has been accomplished by those who have trodden the paths of natural science under more favourable circumstances. His example has given a stimulus to the study of Nature in the districts around him, and he ever takes delight in initiating all who seek his aid into the best means of pursuing the investigation of natural objects."

In botanical study Gibson was hardly less enthusiastic than he was in geological. His contributions to the *Phytologist* and other magazines bear testimony to the extent of his knowledge. Mr. Henry Baines, in the preface to his 'Flora of Yorkshire' (1840), says: "To Mr. Gibson, of Hebden Bridge, the catalogue is under great obligations not only for the free communication of his discoveries, especially in cryptogamic botany, but also for his attention to the general completeness of the work." Whilst the volume was passing through the press, Gibson furnished a list of twenty-seven additional plants, occurring in various parts of the country, which with contributions by Dr. E. Lankester and Mr E. Moore, was printed at the end of the volume. Gibson contributed also to Mr E. Newman's 'History of British Ferns and Allied Plants' information as to new varieties, and stations for previously known ones, which he had discovered. "His wonderful powers of observation, acute discrimination of the slightest difference of form, unwearying industry, and extraordinary neatness in arranging specimens, were such as seldom fell to the lot of one person."

Gibson was an excellent entomologist. He commenced the study of this science in 1826. His acquaintance with it brought him into communication with many of the leading entomologists of the country; and in the course of a few years he brought together a very valuable collection of insects arranged in thirty-four boxes. He, moreover, found time to study somewhat of conchology, and made a large collection of the land, fresh-water, and marine shells of Great Britain. He also possessed a considerable assortment of foreign specimens.

Though not a Lancashire man, Gibson was intimately connected with the botanists of that county, and frequently attended their meetings at Manchester. He took great interest in those meetings, and when the Manchester Mechanics' Institution became the subject of some adverse comments in a Yorkshire paper in 1842, he replied to the editor with some warmth, and said: "The secretary of the natural history department reports that that branch continues to meet regularly once a fortnight, and that several very interesting papers have been read during the year." He adds that "several of the members have, of late, contributed articles to the botanical and entomological magazines—a sufficient proof of their zeal and proficiency in those departments of science. The chemistry class held thirty-three meetings and had ten papers read."

During the last few years of his life ill health, we are told, affected both the spirits and the natural good temper of the man. A fall from a school building, in course of erection, by which several of his ribs were broken, had a very prejudicial effect upon him. He never fully recovered from the injury. Whether it was this which brought about that infirmity of temper under which he undoubtedly laboured, we do not know; but it is said that that infirmity brought him into frequent collision with his fellow-students. At any rate, under his affliction, and the adverse circumstances of which it was the occasion, it is impossible not to sympathise with him. Precluded from following his ordinary employment, and having a large family to maintain, he established himself in a small inn at Mytholmroyd, a few miles from Hebden Bridge, one large room of which he fitted up for the reception of his natural history specimens, hoping that their exhibition there would draw visitors as well from a distance as from the immediate neighbourhood, whose custom would be of some account.

The scheme however failed, and poor Gibson, after removing to a small cottage near the railway station at Mytholmroyd, found himself obliged, under pressure of want, to sell a considerable part of his valuable museum. The part disposed of comprised the geological collection, the birds, and the land and fresh-water shells. The fossils from the lower coal measures of Todmorden, which with other portions of Gibson's collection, had excited the admiration of the *savants* of the British Association (who first visited Manchester in 1842), were purchased for the Manchester Natural History Society's Museum. The French Government, it is said, were desirous of purchasing the entire collection for the *Jardin des Plantes*, but the negotiations failed. It must have given poor Gibson severe pain to part with his most valued collection. It was his last resource. He succeeded, however, in preserving his herbarium of flowering plants, which was complete to within about twenty specimens, and which, after his death, was purchased by Mr. Mark Philips, M.P. for Manchester. It was valued at £75. Gibson also left the large entomological collection before referred to, and his collections of Mosses, Lichens, and marine Algæ; about 1000 specimens of seeds and seed vessels of British and foreign plants, ingeniously mounted on glass, and about 140 specimens of wood sections mounted for the microscope. The collections of seeds, shells, insects, dried ferns, &c. &c., found their way, ultimately, into the Peel Park Museum, Salford.

A curious story is told about the entomological collection, which, together with the other curiosities remaining, and books on natural history, the friends of Gibson were desirous of selling at the best possible price for the benefit of his widow and children. Mrs. Gibson, who had no knowledge of the actual value of the collection, sold a number of boxes of insects to a certain clergyman (who shall be nameless) at a shilling per box. When the fact came to the knowledge of Mr. Binney, who was one of the best friends Gibson had, that gentleman remonstrated with the reverend purchaser, and urged that the boxes should be given up, and re-sold at a price more commensurate with their value. The remonstrance was unheeded. It was the subject of some angry correspondence, and finally the interposition of Professor Sedgwick was sought in the hope that he would make some impression upon the obdurate parson. The Professor, whilst offering an opinion that Mr. Binney had been "rather too hard" upon his opponent, said that the boxes ought, certainly, to be given up and the money returned to the purchaser. This arbitrament was not disputed. Mr. Binney sent the clergyman 45*s.*, and he received the boxes in return. They were afterwards sold by auction, in Manchester, for £45. Mr. Binney informed the clergyman, by letter, of the result of the sale, and said that no doubt he (the clergyman) would be glad to hear it, and to know how greatly he had been mistaken as to the value of the property; but no reply to this was vouchsafed.

With regard to Gibson's early life only the same story can be told as has already been told respecting the naturalists of Manchester. He was born of humble parents (his father was a whitesmith and a Methodist local preacher), and had no education except such as he could pick up in a Sunday school. He used to say that he never attended a day school a single hour of his life. He served an apprenticeship with his father, and became an ingenious mechanic and a most expert wood-turner. At the age of nineteen he married. It was not until he was twenty-five, or thereabouts, that he undertook natural history studies with energy—although for some years he had been a student in a humble way—but, having commenced business on his own account, and having, consequently, more freedom to devote his mind to those subjects, when he began, in earnest, to study geology and botany, he made advances which, considering his want of education, were simply astounding. He had an excellent field for observation, in his own neighbourhood, and particularly in the vale of Todmorden, and from the renown which he achieved, it is certain that he made the most of his opportunities. He was 59 years of age when he died.

ELIAS HALL, OF CASTLETON.

Elias Hall, who was well known, throughout the Northern Counties—particularly in South Lancashire and his native county of Derbyshire—as a most laborious geologist and mineralogist, lives in the recollection of many as a remarkable instance of genius in humble life. Stern self-reliance and indomitable perseverance were the great features of his character. He was emphatically a self-taught man. The works upon which his great reputation as a geologist are founded are, first, "A Mineral and Geological Map of the Coal-field of Lancashire with a Part of Yorkshire, Derbyshire, and Cheshire," coloured stratigraphically—a work which involved much care and labour—and, secondly, sections of the strata across country from the Irish Sea to the German Ocean. The counties traversed by Mr. Hall, in the preparation of these sections, were Lancashire, Derbyshire, Yorkshire, Nottinghamshire, and Lincolnshire; and as they were the result of laborious personal investigation, upon which long experience and extensive knowledge were brought to bear, they are quoted to this day as models of accurate detail. The map was dedicated to Professor Sedgwick; and an introduction to it, published subsequently, to which was appended "A Medico-

Topographical, Geological, and Statistical Sketch of Bolton and its Neighbourhood" by Dr. James Black, was dedicated, "as a trifling proof of respect for a life devoted to the pursuit and improvement of a favourite science," to Professor Buckland.

In the preface to this latter work the author trusts that his map will be received by the public as the mineral is received by the smelter, "with a view to extract the purer metal for his remunerating purposes." With great truth he observes: "In all inquiries of the nature now proposed, it is highly necessary to exercise the greatest caution lest the ardent imagination of the geologist should lead him into error, by causing him to draw false conclusions. Results propagated to the world upon data not sufficiently accurate must ultimately tend to prejudice the author in the scientific world, as well as to create confusion and generate numberless errors." Fully sensible of the necessity of this caution, he has, in constructing the map and preparing the introduction, carefully abstained from all theoretical opinions, and confined himself strictly to a record of well ascertained facts the correctness of which he has verified by actual observation.

An introduction to the sections, prepared by the author with the assistance of Mr. Francis Looney, giving an account of the fossils discovered in the several strata, was also published. Mr. Hall afterwards commenced the preparation of a geological map of the Midland Counties, and he was engaged upon it to within a short period of his death.

In addition to the works already enumerated, Mr. Hall completed several carefully executed models of portions of the earth surface, including the Peak of Derbyshire and parts of the Lake District of Cumberland and Westmorland, in which with the aid of colour the stratification and other geological details are indicated with admirable clearness and accuracy. Two of these models were, at the instance of Sir Joseph Banks, purchased by the trustees of the British Museum, and are now deposited in the geological gallery of that institution; another —a model of the district around Manchester—is preserved in the Natural History Museum of that city.

Mr. Croston, to whom we are indebted for some of the above facts, in his work, 'On Foot through the Peak,' says that when the celebrated French naturalist, M. Faujas St.-Fond, visited this country, Mr. Hall accompanied him through Peak's Hole, and showed him the wonders of that remarkable cavern; and some fifty years ago, when Mr. Farey was collecting information for his 'View of the Agriculture and Minerals of Derbyshire,' he rendered him considerable assistance, revising and correcting those parts of his work which relate to the stratification and minerals of the county. "Mr. Hall," he adds, "may fairly claim to rank as the father of geology in Derbyshire. He first directed attention to the subject at a time when geology, as a science, had made but little progress, and, in this country, was comparatively unknown, and he continued his investigations with unceasing application for more than seventy years. We remember him, in the later years of his life, a fine, hale, and hearty old man, with an energy and restless activity remarkable in one of his advanced age; plain, homely, and unaffected, with a cheerful and social disposition, and a kindness of manner that secured for him the friendship of all with whom he came in contact. He was ever ready to afford information, and to communicate unreservedly the results of his investigations to those who desired to possess them. He retained his mental faculties to the last, and died on the 30th November, 1853, in the ninetieth year of his age."

Mr. Hall described his business as that of a "mineral surveyor," and he advertised for sale, at his residence at Castleton, his "models of the strata of several districts, and the accompanying minerals." Some years after his death a few of his admirers in Manchester and the neighbourhood, caused a neat headstone to be placed over his remains in the Castleton churchyard, bearing the following inscription:—

"IN MEMORY OF
ELIAS HALL, THE GEOLOGIST,
who died on the 30th day of December, 1853,
AGED 89 YEARS.

"Born of parents in humble life, and having a large family to provide for, yet he devoted himself to the study of geology for 70 years, with powers of originality and industry rarely surpassed.

"To mark the last resting-place of one who had worked so long and so hard for the public, a few of his friends and admirers, living at a distance, have placed this stone."

TWO SCOTTISH NATURALISTS IN HUMBLE LIFE.

The "Mad Baker" of Thurso.

There died, in the year 1867, one of the most remarkable men of modern times, and one who may fairly be said to have left his mark upon the scientific achievements of the century. His name was Robert Dick.

Although a native of Fifeshire, Robert Dick lived during the greater part of his life in the town of Thurso, whither he removed when young. He learned to be a baker, and commenced business on his own account; but, from his peculiar disposition and habits, he was never able to establish a very large or lucrative business.

During his apprenticeship Mr. Dick began to manifest those extraordinary mental abilities which were in mature years fully developed. He would then spend even more than his spare hours in local explorations, and every work on botany and entomology was eagerly borrowed or acquired, and was read and studied with the greatest avidity. But it was when he became a journeyman, and especially when he arrived at the position of being his own master, that he devoted himself with the most singular earnestness to the study of science, spending, during his life, many nights in the open air, and being on many occasions several days and nights engaged in these investigations in the district which in the end brought him into possession of a museum of fossils and botanical and entomological specimens which has been the admiration of the multitudes of *savants*, from Sir Roderick Murchison downwards, who have been privileged to see it. Among the people of Thurso and neighbourhood Mr. Dick was long looked upon as partially insane. But as time rolled on opinions gradually changed. By-and-by it began to be whispered that men of great influence were visiting the mad Thurso baker; and when it was found that in the meetings of the British Association for the Advancement of Science he was named as one of the highest authorities in the world on certain scientific questions, and that even Sir Roderick Murchison had been sitting at his feet and receiving lessons from him—some of them characteristically drawn on the walls of his workshop and his implements of trade—the opinion changed, and Thurso people took pride in naming the great scientific baker of their town.

It was during his entomological and botanical explorations that Mr. Dick began to cultivate a taste for geology. By-and-by he became as deeply in love with it as with other sciences; and in the end he acquired a wonderful amount of geological knowledge. He communicated to the late Hugh Miller, Sir Roderick Murchison, and other geologists the result of his investigations, which have had great influence on the present knowledge of the science.

The result of his long and wonderful excursions in the North of Scotland, and of his extraordinary painstaking investigations, was that he accumulated specimens, which in the end formed a most valuable Museum of Natural Curiosities, which is now the property of an association in Thurso. In British botany, his name will always be associated with the discovery, amongst other treasures, of the Northern Holy Grass (*Hierochloe borealis*), which, until he found it growing near Thurso, was believed to have been peculiar to the North of Europe.

Mr. Dick's devotion to science, coupled with his personal manners, always kept him a poor man, and the result was that he died steeped in poverty. No relative attended his funeral. He was buried among strangers, and in the neighbourhood of the localities where, when others were enjoying their repose, he spent many a night of patient and plodding investigation, and succeeded in giving to the district a fame in various departments of natural science which but for him it would perhaps never have enjoyed.

Thomas Edward, of Banff.

Thomas Edward, a journeyman shoemaker—now, and for some years past, also curator of the Banff

Museum—is another remarkable instance of devotion to natural science; and his whole life shows how much a man with the slenderest of means may accomplish, if he has but enthusiasm and natural energy combined.

Mr. Edward, from his very earliest years, had a strong love for all pursuits connected with natural history, and he has prosecuted his researches in all its departments. He is not merely an accomplished zoologist and botanist, but also a good conchologist, and latterly he has devoted a large share of his attention to marine objects.

His early researches were, of course, made under the greatest difficulties. He had scarcely any books; and he had thus, in most instances, to gain his knowledge of the characteristics and habits of the various animals and objects mainly from personal observation. Often, when he thought he had discovered a new species, it was found that the animal was known before. But his habits of close observation, persistently pursued day after day, could not fail to bring to light new species from time to time.

From his want of systematic works on the various branches of natural history, Mr. Edward has been obliged to place his discoveries in the hands of others; but his discoveries in this way have been very great. He has been fortunate enough to present to the various departments of natural history at least thirty new species. These consist of Annelids, tunicate Molluscs, Sponges, shrimps, and the smaller species of Crustacea. Besides redeeming one from oblivion, he has also brought to light two entirely new species of fish. With regard to the long-lost *Gadus argenteolus* he thus wrote, in 1864:—

"In the year 1808, Colonel Montague, a well-known naturalist, found a small species of fish on the Devonshire coast. This he described in the second volume of the 'Memoirs of the Wernerian Natural History Society,' and called it the silvery gade (*Gadus argenteolus*); no figure was given, at least so far as we have ever heard, or have been able to ascertain, but simply a description. But by some means or other it was considered by many that the Colonel had been mistaken, and had described a fish as new which had been previously known, by omitting to notice some of its distinguishing features. This in process of time grew into a settled belief, from the fact that the fish was never again, even so much as once, met with. Had it in reality been a distinct species, where could it have gone to, so as never again to be found nor seen? . . . . Years rolled on, and the Colonel himself passed away, but still his little fish was never once more heard of, save by writers on these subjects, who gave meagre accounts of it, abridged, no doubt, from the original one, all the same time doubting whether it was or was not a species. And to crown all, it was at last entirely swept from creation, as it were, and blotted out altogether from the page of history, as a thing that never had been and as having no right ever to have been there. This will be better understood when it is stated that it is wholly excluded by that veteran and accomplished ichthyologist, Mr. Couch, from his new and great work on 'British Fishes.'

"But what will the *savants* in these matters say when they come to learn that it has been again met with, and that, too, at Banff, far away from the coast of Devonshire, and just fifty-six years, too, after Montague found the first? However wonderful, and however incredible, this may appear, it is, nevertheless, an undeniable truth, for we were fortunate enough during the first week of last month (October) to take several specimens, both big and little—that is, young and old. Of this there is no mistake. But, by way of proof, if anything of that kind were wanting, or at all necessary, we may just mention that, having sent specimens to Mr. Couch, that gentleman, after thanking us in the most enthusiastic manner for the unlooked for and unexpected favour, and having alluded to the great importance of the discovery, says:—

"'Although too late now to be noticed in its proper place, that part of my work being already out, I will, however, most assuredly insert the species, along with some other matter, in a supplement, and shall in particular mention you as its re-discoverer.'

"Some writers say that this species (that is, the silvery gade) and another little fish called the mackerel midge are similar, the only difference, according to them, being that the one has two and the other four barbules on the head. These are individuals who had never seen the fish, or, having seen the one, had never seen the other. We have seen both, have taken both, have had both alive in the same vessel beside us at the same time, and can say of a truth that there is a most marked and striking difference betwixt the two. They are not at all of the same form or make. . . . But enough, especially as the species will be faithfully described by Mr. Couch, with this difference, that the silvery gade will henceforth be called *Montague's midge.*"

Mr. Edward adds:—"We may just mention, by way of a sequel, that this genus (British) of little fishes, designated with the English appellation of *midges*, from their small size, and containing three species, are now authentically known to be inhabitants of the Moray Firth, all three, both young and old of each, having been procured here, a circumstance which, perhaps, can be said of no other single district but our own. This, not so much for the lack of the fish themselves, as the want of searchers for these things, for we cannot allow ourselves to think for a single moment that they would or could be found in so widely distant localities as Cornwall, Belfast, Devon, and here, and not to be met with at intermediate stations. Such a thing appears to us to be one of those affairs usually called *impossibilities.* Let those, then, who live on the coast, and have time and a mind to these things, or whether they have time or not, if they have the will, there is no fear of the other; let such, we say, look better about them, and we doubt not but they will find these little gems, as well as other rarities of a similar and kindred sort."

Mr. Couch has been indebted to Mr. Edward for many notes for his valuable work on Ichthyology. To Mr. Bates' work on the crustaceans, Mr. Edward has contributed, possibly, as much as all his other correspondents put together. Mr. Edward's devotion to his favourite pursuit at one time seriously endangered his health. The reader will not wonder at such a result when we say that it was his custom, for nearly five successive years—from 1840 to 1845—to spend most of the nights in the open air. Indeed, he might during that period be truly said never to have been in bed save a small portion of the first and last days of the week. It was his daily practice, on returning from work at night—and shoemakers then worked from six A.M. to nine P.M.—to bundle on his collecting apparatus while he was taking his supper, and start with a portion of his meal in his hand. His collecting apparatus comprised insect boxes and bottles; and he had, usually, a botanical book, and, besides all, a gun. So accoutred he scoured the country, or as much of it as he could, before darkness compelled a pause. When he could no longer "observe," he dropped down by the side of a stone, or dyke, or tree, whichever came handiest. Here he lay, sat, or slept, as the

N

case might be, till daybreak enabled him to resume his searches, which were continued till he had to return to his daily labour. It was no unusual circumstance, when he had wandered too far, and come upon a more than usually attractive spot, that he would strip himself of his gear, gun and all, which he would hide; and, thus lightened of everything save his specimens, would take to his heels, and run at top speed to be at his work at the proper time. Weather never daunted him; and his neighbours used to say, "It was a stormy night that kept that man in the house." During the years we have mentioned, and for four or five years before 1840, Mr. Edward collected over 2000 zoological specimens. Those were carefully preserved and arranged in cases. The cases, nearly 300 in number, were all made by his own hand, his only tools being a saw and his shoemaker's knife. The cases were neatly painted and prepared, and had glass fronts.

In addition to his natural history studies, Mr. Edward paid considerable attention to local antiquities, and published an account of the Kjokkenmoddinger ("kitchen middens") at Boyndie, upon the remains found in which he was enabled, by his zoological knowledge, to throw considerable light. He was the author of some 'Notes on the Antiquities of Banff.'

The *Banffshire Journal*, in giving a brief account of Mr. Edward's labours some years ago, said:—

"We wish we could say that the results of his labour had added aught to his pecuniary resources.

He frequently exhibited the collection at the markets in town. But in an evil hour he was advised to take the collection to Aberdeen, and exhibit in that city. The transport thither—it was in 1846, before the days of railways—was not cheap. He took a shop in Union Street at £1 a-week, and advertised liberally. But the Aberdonians had greater attractions than the natural history specimens of a local collector. A few dropped in; but the receipts were miserable, while the expenses were heavy; and at last, after exhibiting for about six weeks, he was obliged to succumb; and, to save himself, and meet the outlays he had incurred, he sold off the entire collection. The expenses paid, he had scarcely as much left as paid his way home. So went his ten years' labours and hard earnings." Mr. Edward made a second collection, not so large or complete as the first; and with this he was obliged to part in a period of ill-health. He began a third; but for the same reason it, too, was partially disposed of.

Mr. Edward was, many years ago, elected an Associate of the Linnæan Society in recognition of his services to science. In that society's journal there appeared (Zoology, vol. ix.) "Stray Notes," by him, on some of the smaller crustaceans; and in vol. x., "A few additional particulars regarding *Couchia Edwardii*, the name of one of the two species of fish which Mr. Edward discovered in the Moray Frith, and which was named, specifically, in honour of the discoverer.

The local offices which Mr. Edward holds—those, namely of curator of the Banff Museum and librarian of the Banff Literary Society—are both offices of small emolument.

Mr. Edward, upon being informed by the author that the present work was in progess, wrote him the following letter. Although partaking of the nature of a private communication, it is here given *in extenso* and as it left the writer's hands, because it represents as clearly the character of the man as a photograph would represent his features, and shows, in his own language, how passionately devoted he has been, during a long life, to the study of Nature:

"My dear Sir,—Yours of the 28th April reached me in due course.

"You are quite welcome to use any information you may have, or may yet acquire, concerning me, and the way best suiting your purpose. I would advise you, however—seriously advise you—first, to read, if you have not done so already, the *Zoologist*, the *Naturalist*, the *Banffshire Journal*, of bye-gone years, Mr. Bates' recent work on the British Sessile-eyed Crustacea,' and the 'Story of a Rural Naturalist,' in *Chambers's Edinburgh Journal* for Saturday, July 31, 1858. Do this before you attempt to say anything about me in your forthcoming work. And even when you have done all this—supposing you do it—and unless you are gifted with something far above the average capabilities of our race, you will fall miserably short of my true character, at least, so far as relates to my unbounded adoration of the works of the Almighty—not only these here on earth, but all those wondrous and countless millions of orbs which roll both near, if I may so speak, and far away, in the endless immensity of space, the Home of Eternity.

"Every living thing that moves or ever lived (I allude to fossils), everything that grows, in short, everything created or formed by the hand or the will of the Omnipotent, has such a fascinating charm about it to me, and sends such a thrill of pleasure through my whole frame—especially when handling or viewing anything new or rare—that to describe my real feelings on such occasions is utterly impossible. Language has not the power. Such being the case, you must just do, unless gifted as already hinted, as others have done before you—guess; for I have never seen anything yet concerning me, either way, that has ever come near to the truth. Fiction! Pooh!—nothing! Why, if the truth could be realised, and told, it would put the wildest and most fictitious story ever published to the blush.

"Willing at all times, when able, to give every information in my power relating to these things, and my own career, and without a single thought as to how such may be appreciated, I will do what I can to aid you in the way you suggest.

"In the first place, however, I would beg of you

to bear this undeniable fact always in your mind, or before your eye, that I am only a *journeyman shoemaker*. Do not, I say, forget this truth. Never mind though I am *this* and *that* of several learned societies; still I am but one of Crispin's poorer sons, and will, no doubt die such. I have often heard people wonder why I had never got above the old stool, as they thought I might have done; and, perhaps, you will wonder too, not knowing better.

"It is an easy matter—quite easy—for anyone having education and a friend or two to get on through, or even to rise a bit in, the world. Not so with the poor fellow who has none of these. I had neither.

"Put to work when little more than six years of age, what learning could I have had by that time? And, besides, the school proved such an abominable prison to me, that I could never endure it. The consequence was that I oftener—much oftener—played the truant than the faithful scholar. Besides, when I did attend I was generally such a source of dread and terror to the whole school, with my beasts, as I always had lots of them about me, that I was invariably put out before the proper time. All the scholars delighted in my butterflies, but few, or none of them, liked to be stung by bees or wasps, or bitten by beetles, or to have worms or horse-leeches crawling on or about them. I understood well enough the meaning, in all its bearings, of the *thong*, with its five or six longish points; but the often, and no doubt earnestly repeated advice, or request, 'Now do not bring any more of those dangerous things here again,' I either did not comprehend, or I forgot to act upon the friendly suggestion. The neglect of this simple precaution in the end, if I remember rightly, caused my final dismissal from all the schools—three in number—I had the sad mortification ever to be put to.

"Now, as to friends. Well, I may say, I have none! Some are very fortunate in this respect. I cannot say that I have been. Some are very good at making these things and first rate at retaining them when made. I have never, as yet, been able to find out the secret of the one nor the mystery of the other. How, then, was it possible that I could have risen above the lapstone and the old ricketty soap-box which I converted into a shoemaker's seat, and on which I still sit?

"But, besides being ignorant and friendless, I had not even so much as a single book on the subject, from which I could have got the least information, for many years after I had begun my researches as a naturalist, all my findings being made, as it were, by mere chance. But this very privation proved one of the most gratifying and inexhaustible sources of pleasure to me, which anyone could enjoy. I knew almost nothing of the lives and manners of the various animals I was about to seek, and only a little of where a few of them could be obtained. So, having no books, and no one to direct me, and being anxious to learn—this led me, despite my eternal craving and all but ungovernable longing, to possess the creature first, if by any means possible, to make myself as far as I could acquainted with its various traits and habits, before I made any attempt to kill. And this has been my custom, all through life, when at all possible. By this simple means, I have been able, in a measure, not only to store my mind with much that no book but this Book of Nature could impart, but also to enjoy many a pleasant hour which I could not have had in any other way. As a natural consequence, I sometimes lost the object by this procedure; but if I had learned anything concerning it that I had not known before, it helped to lessen my grief, as that was so much gained—mentally.

"I have often been asked what made me a naturalist. When first asked the question, I was completely dumb-foundered; I had no notion that a naturalist could be *made*. What!—make a naturalist as you would make a tradesman, or such like, so as to make his living by it? The very idea of such a thing seemed preposterous in the extreme to me. My answer to all who have asked me has invariably been—I could not tell, for the very plain and simple reason, I did not know.

"But, I suppose, it has been the same internal impulse which prompted me, when only about four months old, to leap from my mother's arms in the vain endeavour to reach some flies creeping on a window near to which she stood, and which would most likely have proved fatal to me had it not been for my long dress to which she clutched, and thereby saved me from falling to the ground. This unseen something—this double being, or call it what you will—inherent in us all, whether used for good or evil, which stimulated the unconscious babe to get at, no doubt, the first living animals it had ever seen, having grown in the man into an irresistible and unconquerable passion, goaded him on, whether he would or not, and engendered in him an insatiable longing for, and earnest desire to be always amongst, such things. This is the only reason I can give for my becoming a lover of Nature. I know of none other.

"But this I know: that, ever since I can remember, I always delighted in every living thing I saw, and never missed an opportunity of taking all home I could lay my hands on, even although I frequently came to grief for my pains. These consisted chiefly in my earlier years, of beetles, lots of flies, bees, wasps, worms, snails and mice. Later on, I had rabbits, hedgehogs, guinea-pigs, frogs, toads, horse-leeches, fish, crabs, a black rat, and a few birds, mostly sparrows. Some of these—especially the insects—frequently got away, and caused rather serious and angry commotions, not only amongst my own folk, but also amongst the neighbours. So far were the houses, at times, overrun with what the people were pleased to call that '*Confounded laddie's venomous beasts*,' that, on several occasions the neighbours complained to the landlord, and he, at last, threatened my parents with immediate ejectment if they did not at once put a final stop to the nuisance. This in some measure had the desired effect; but I managed to retain and keep, in one way or another, the larger portion of those I still had; and, although I was almost daily found fault with and not unfrequently threatened with corporal punishment, or something like it, and oftentimes obliged, reluctantly enough, to put away what were termed the most 'dangerous' and 'loathsome' (unmeaning terms to me), still—and I remember it well—it never had the least effect in cooling my ardour or lessening my affection for these things in any way.

"I was nearly two years at a tobacco work. From this I went to a factory, and, after being there about the same time, had the unspeakable misfortune to go to the business I still follow—that of a shoemaker. The factory period, I do think, was the sweetest episode in my whole boyish life. Situated, as the mill was, in the centre of a beautiful valley, and almost embowered in tall and luxuriant hedges of hawthorn, with watercourses and shadowy trees between, and large woods beyond, all teeming with Nature, it appeared a perfect paradise to my young fancy. And, although I had to walk fully two miles to and fro, each morning and night, still it was delightful to me, especially in summer. It was, truly, a happy time, so many nests, so many wild flowers—so much of everything—I did, indeed, leave it with sincere regret.

"The shoemaker to whom I had the sad misfortune to be bound, of all men in the world, was surely the very worst I could have met with. He cared not for the works of Nature himself, and hated it in others. Many, and often hard, were the tousels we had on his seeing that I would not submit quietly and allow him to rifle my pockets whenever he chose, that he might either dash the contents to the floor, to be trod on, or cast forth into the street to the dogs. And many a blow have I got from the inebriate for my defending and trying to keep my own. He would 'stamp the fool out of me,' he used to say. But he did'nt. The 'fool,' as he called it, had been too firmly and deeply implanted, and by a higher and wiser hand than his—poor soul!—for him to be able ever to root it out. Though I did all I could to avoid giving offence, still little things would crop out now and then, in spite of me, and always bred mischief. But the climax came at last. After knocking me down one day with a last, he seized me by the neck and breast, dragged me to the door, and then, with a terrible imprecation (at which he was very good), threw me out—and all because I had resisted his taking my bonnet, that he might have his vengeance wreaked on three young moles it contained.

"So terminated my apprenticeship, after being three years and six months at the trade. A few days afterwards, I began to work for another master; and now as a journeyman, though only about fourteen years of age.

"Having now, comparatively speaking, my time all my own, and having little or no restraint to hinder me from following after my favourite object —the magnet of my very existence—I need scarcely say that no opportunity was lost, all being taken advantage of to the fullest extent. But it was not until I became settled in Banff that I began, I may say, in real earnest. By that time I had contrived by one means or another, to learn how to stuff a bird and to pin an insect down to a board. With this large stock of knowledge I commenced, in 1837, to form what may be called my first collection. I had a short time previously bought an old flint-lock gun which cost me *four and sixpence;* and so ricketty was it that I had to tie the barrel on to the stock with a piece of twine. Such was my *gun*, but I had no *dog*. My powder I carried in a horn and my shot in a paper bag. This, with a few small bottles and some boxes that I put into my pockets, was my whole equipment.

"Shoemakers in those days had to work much longer daily than now, and, even then, were only able to earn a very scanty livelihood. The young men in Banff all wrought in their employers' shops, and, though paid by the piece, had stated hours for commencing and ceasing work—six in the morning and nine in the evening. At that time, too, the work was much heavier—stouter—so that less could be done with even harder labour, whilst the wages were most miserable. I am not so sure about it now; but then, before, and for many long years after, it was in every particular, and in every sense of the word, the most wretched trade that any human being could have been put to.

"Such being the case, I found it indispensably necessary to husband well both my time and what little money I made, so as to make the most of the one and the best of the other. And in order the the better to accomplish this I made it a maxim to never spend a single moment in idleness, or at least which I did not try in some way or other to improve: lived soberly and never spent or squandered a farthing in profligacy, or anything pertaining thereto. Although I sometimes took a whole day, or part of a day, the most of my researches, in summer, were made betwixt dropping my usual work at night and resuming it in the morning. To make the most of my time in this way, I generally, having equipped myself, took my supper either in my hand or my pocket, went out, and did not return until it was time to commence work the next morning. I searched about as long as light lasted, then lay, or sat down beneath a bush or by the side of a bank, or stone, or tree, or whatever came nearest; and was up again whenever the first streak—and sometimes before it—of dawn appeared. If it rained, I dropped into any shelter I could find. If a hole, I went in feet foremost, keeping my head out in case of casualties, and to look about me. Two such places I frequented, and they proved excellent retreats many times to me on wet nights. If it rained heavily I drew in my head, and was then as snug as a thief in a mill or a snail in its bucky. I have shot such visitors as owls, and on one occasion a badger.

"My rounds extended coastwise about seven miles in one direction and six in another, and from five to six inland. I had thus three distinct courses, as it were; and although I could not visit them all in one night, I usually managed to go over them all once, if not twice, a week.

"Thus, from 1840, at which date I commenced my thorough out-and-out night-work, till 1855, and during the greater part of each year, I very seldom slept at home, or in a house at all, unless a few hours on Sunday morning, and again on Sunday night. This latter, however, was much oftener spent on a chair, having first stripped myself of my Sunday, and donned my working, clothes, to be the readier to start by, if not before, the peep of day. Having wandered too far of a night, as I sometimes did, or been otherways detained, I usually on such occasions divested myself in the morning of my paraphernalia—except what specimens I may have had—and, having rolled my gun and all well up, hid the whole in some convenient place till next night. Thus disencumbered I ran home, in order to be at my work in time. When I met any one I always pulled up—there is no saying what may have happened had I not done so—but, as soon as we were parted, away I flew again. I once ran in this way fully three miles in twenty-four minutes.

"When I asked away for a few days, as I occasionally did, to extend my investigations farther than I could possibly do otherwise, I always took as much food with me as I thought would serve for the time, and adopted the same plan when night came down as I did when on my shorter excursions —namely, the earth for my bed and the canopy of heaven for a blanket. Sometimes, however, and when they came conveniently, I took up my quarters in any disused building, in ruins of old castles, and such like, or in churchyards, as affording a little more comfortable shelter than a hard stone or bush could do. My only objection to such places was that I had a much larger number of visitors than were to be had outside, such as polecats, weasels, rats, mice, snails, and hosts of night-prowling insects, with hordes of slaters and centipedes. These latter, however, were as nothing compared with the former, as those of them which remained, and were seen in the morning, could easily be rubbed or shaken off; whilst those that had crawled up my legs or arms generally dropped away during the day. But think of having a polecat or a weasel sniff-sniffing at your face as you lie sleeping, especially if you have got any idea of their blood-sucking and throat-cutting propensities, either from reading or having been fortunate enough to have been told such horrible things!—or even of having a ratten or two, or perhaps more, tug-tugging and — bold fellows!— attempting to drag away your larder, stealing from under your very nose all you have on earth to sustain your own life. Yet I have met with all this and more. But such intrusions did not always prove an annoyance to me. They oftentimes turned out a windfall—that was, when the intruder came within my grasp and was found worthy of being retained.

"Amongst the most remarkable nights I ever spent in this fashion, and one which I shall never forget, was in a churchyard. I had just crept beneath a tombstone, which was supported on four low pillars, when one of the most terrific thunder-storms I had ever witnessed burst forth. I did expect a wild night as I had seen it brewing all the afternoon; but not such as it turned out to be. Such lightning, or rather bolts—huge bolts of liquid fire; such flashes, again and yet again! what red and yellowish fiery-like streams; and what a disagreeable, suffocating sulphurous smell each flash left in its trail! And what thunder! What continuous peals, and what incessant crashing and rumbling! Now, louder and louder, nearer and nearer. Then, what torrents of rain and snow! What drops, and what hail—or rather triangular lumps of ice! It was altogether a sublime and glorious spectacle, and one but seldom to be seen in this country. Yet again! what flashes: how blinding and painful to the eyes. What crashing and tearing: and—how near!

"What a night it was! as if the whole universe were being torn asunder, and the rough and mighty fragments hurled precipitately down some steep and rocky declivity. How terrible!

"Afraid? No; I can scarcely say that I was. My situation was a rather peculiar one, I own; and I dare say not many would have cared to occupy it at that time of night, especially on such a night. Few would have liked to forego the pleasures and comforts of home, even although it was to behold Nature in one of her grandest moods and to listen to the music of her artillery. Instead of being afraid, I rejoiced in the fact, for the night had unspeakable charms to me. My only regret was at being so low down. Had I been higher, and closer to the tumult, that I might the better have beheld the working of the elements, I should have been much better pleased. But that was not to be. And, as most things here have a termination, the aerial strife at last ceased; and then the wind, of which there had been none during the thunder, began, lightly at first, but rapidly increasing, it soon blew a hurricane. Though I had no fear of the thunder I had a little of the wind, as it now was. At a few yards' distance there stood, by itself, the gable of an old church, towering right over the very grave where I lay. I knew that it would be rocking and swaying to and fro, for nothing of the kind, as I thought, could resist such a tempest. I knew likewise what the inevitable consequence would be were it to give way. True, I had a stone above me, but, for all that, I would much rather not have risked such a lump of masonry coming slap down over me.

"It was still dark, and I knew of no other shelter near by. There were houses close at hand; but though I would not have cared if I had been looked upon in the light of a resurrectionist, yet I had no wish to be considered, or, it might have been, taken up for, a burglar; so I remained where I was, dreading the worst, as we mortals too often do, but hoping for the best.

"Happily, however, the wind, after roaring and bellowing fearfully for fully an hour, began to moderate, and ultimately ceased. My misgivings concerning the old wall having also vanished, I desired now, as morning was approaching, to betake myself to a few moments' repose; and—vain wish!—I had just begun to doze a little, as it were, when I was again awakened by the most weird and unearthly moanings—or what seemed to me a double series of moanings—which it is possible to conceive. They were quite near. The noise continued until it became a sort of stifled scream, or, more strictly, a perfect Babel of screams. Babel! Babel! it went, commencing in a low key, but ascending gradually until it reached the highest pitch; then, descending, it would end in a gurgling and vicious moan. Whatever it was, it appeared to me to be double. Had I been near a large fish-market, instead of in the home of the dead, I might have guessed the cause. As it was, I could not.

"Up it went again; this time like the howling of a mad dog, or dogs, long drawn out. Now a lull, then a rushing to and fro amongst the grass for a few seconds, and all was quiet once more. I was about to rise to see if I could make out what it could be, when something light in colour dashed past with a flash, closely followed by another something, but dark or blackish. Ho! thought I, here is at least one thing I did not believe in—*a ghost.* Yes, a ghost, and no mistake, and the devil after it!

"I had more than half made up my mind to rise and join in the pursuit; but, then, however much I might have longed to possess the first as a specimen, the other—aye the other!—stood in the way, in so far as I remembered, on second thoughts, how we are advised—and I suppose it's all right—to flee from rather than fly after him. So I, at least for once, turned my back as it were on scientific research for fear of the devil!

"Having thus, wisely or unwisely, given up the idea of meddling with other folks' broils, I again shut my eyes, in the hope yet of getting a little rest, and had slept a short time when something brushed rapidly—swish, swish!—over and over my legs, and aroused me. Looking up, I was not a little surprised to see two *cats*—a lightish one and a dark—enjoying themselves right merrily, flying in and out, below, over, and round the stone. In, below, over, and round they went; and when, as I thought, they were tired, both came, very jocose-like, and seated themselves close beside me. I did not care to disturb them so long as they kept sweeping across my legs, but I certainly did think this was coming the joke rather too far. On my intimating this idea to one, they both 'bolted,' and much quicker, I can affirm, than they came.

"As the dawn by this time was beginning to appear, I, too, left the grave, to resume my labours, old horse like, stiffly at first, being somewhat wet and cold; but this wore off as my joints warmed.

"Thus ended the night's vicissitudes, commencing with the thunder and terminating with the ejection of the devil and the ghost from my side, but leaving the old and venerable pile, with its little bell, still standing, despite the blast and my misgivings. I was glad of this, for more reasons than one. I like to see these hoary and time-honoured relics of a long past age: they always do my heart good.

"But though I made it a practice during my nightly rambles to return in time to my work next morning, I found it occasionally both necessary and advantageous for my purpose, though not for my wages, to deviate a little from the rule, being one of those who, having once set their minds on anything never relinquish it until the object of the pursuit is either obtained or hopelessly lost. No wonder, then, that I should have got myself lost now and then for days, and sometimes for nights.

"In one case, a little stint cost me two days and a night before I procured it. Although occasionally within a mile or so of home, I never once thought of visiting it, though I had no food the whole time. So intense was my desire to have the bird, that I never thought of the one, nor felt the want of the other, until success crowned my perseverance. I slept that night—or rather, I lay—where I had slept before, amongst the shingle by the sea shore.

"At another time, two geese, the first of the kind I had ever seen, took fully six days before I managed them. I saw them first upon a Sunday afternoon, and by next morning I was at the place—a piece of water close to the town—before daylight, but (if I exclude myself) no geese were there. In about an hour, however, they—that is, the winged species—arrived, and alighted within a short distance of where I was sitting. Had my plan been that commonly in use among gunmen I might have secured both birds that morning; but I wished, first, to see a little of their ways of working. Satisfied to some extent on this score, I then, and not till then, thought of possessing the birds, and was fortunate enough to kill one. I now wished to have the other; but found this a rather difficult matter. Though it did not leave, it became so extremely shy that it could not be approached. It was only by using the utmost precaution, and having recourse to several, and some very curious, stratagems that I at last succeeded in shooting it. This was on the Saturday; but I did not procure it until next day. This took place in the month of March. I always went home at night.

"I generally contrived to preserve my specimens during my meal hours, and what other time I could spare during the day; and, in the long winter nights, I arranged and put them all into cases. The whole of these I made myself, and with no other edge-tools than a saw and my shoemaker's knife. I papered the inside, painted the out, and put the glass into the fronts all myself. Not being able to afford both fire and light every night, I always put out the lamp when engaged upon anything that I could do without it, and continued my labours by the light of the fire. I also, and not unfrequently, during winter took my work from shop home with me at night, and wrought at it instead of going to bed. This was done in order that I might have more time during the day to go after the sea birds, and such like, that visited us at that season. Moonlight nights did very well for some species and some localites, but not for others. It was quite a common remark of my master's: 'Give Tom the stuff of a pair at night, and if he has any of his tantriffs or whims in view, you are sure to get them from him in the morning *made*.'

"By the summer of 1845 I had thus accumulated and preserved nearly 2000 specimens of the various creatures found in the neighbourhood—amounting to about half that number in species—consisting of quadrupeds, birds, fishes, insects, reptiles' eggs, shells, crustacea, starfish, worms, zoophytes, corals, sponges, and fossils. Some were in bottles, but the great majority were in the cases already spoken of, and which numbered about 260.

"I now thought, being in want of funds to enable me to prosecute my investigations further than I had as yet been able to do—and with the view of realising a little for that purpose—I would exhibit my collection on our annual feeing market-day. With this intent I took a large apartment in a house situate in the street in which the market is held; and prepared accordingly. Well, the day came and passed; and although I did not make so much as I would have liked, still I had a few shillings after paying all expenses; and this was gratifying. I had two very great drawbacks: I was high above the street on which the fair stood, and I had no music.

"Now, music is the very soul of all market shows—perhaps of all shows—and the rougher and wilder the better, as it seems then to have much more influence in drawing people's attention than when otherways. And the more you yell yourself, the louder and more unearth-like your squeal, and the more you can befool yourself, decently or indecently, all is the same. The more expert you are at gullibility you are sure to have more of the crowd to patronise your *sarsaparilla,* or whatever other curious name you may give it; and the more unmeaning the better, as it will draw still more of the wonderstruck and admiring bipeds. There were several shows in the market that day, all having these accompaniments in full force, or I might have done better.

"By next season I had added considerably to my collection. I had now over 2000 specimens and 300 cases; so, an out fair-day coming round, I tried it again, and did better. I would now try Aberdeen.

"On the morning of Friday, July 31, 1846, I accordingly left Banff with six large cart-loads (there were no railways here then), and arrived in the 'hard city' on the evening of the following Saturday. I arranged my specimens on the Monday after, in a large shop I had previously taken for that purpose, in Union Street, and opened on the Tuesday—with what success a few words will explain. At the end of five weeks I found myself deep in debt—by the end of another my collection was gone—all, all gone!—gone for the paltry sum of *twenty pounds!* aye, glad I was to get even that. It was a bitter pang and heart-rending struggle to me to part thus with all my specimens for a mere trifle, or even for money. But what could I do? I had no one to help me. The discerning and appreciating public had already done what they could, and intended to do. I had received two letters from my master—the first requesting me home; the other (I had only sought away three weeks) saying that if I did not come immediately he would be under the necessity of engaging another in my place.

"I had a wife and five children depending upon me. What, then, could I have done?—or what else could any one have done, situated as I was—threatened with loss of work (and no wonder, being so long away), and with want and misery staring my family in the face? Well, to prevent the one, as far as possible, and to keep the other, if I could, I did at last what I did; and nothing else on earth would ever have compelled me to part with my specimens.

"I learned when it was too late that, however well and highly versed the inhabitants might have been in other matters, the masses in general were in a deplorable state of ignorance regarding the works of Nature, and seemed to have little or no taste for anything of the kind. I likewise found out that, for whatever special purpose I had been created, I at least had never been made and sent into the world to be a *showman.*

"I left Banff, as already stated, with six large cart-loads, and returned within a few weeks without a single specimen, and all but penniless.

"But it was not, I may say, until I again came to enter my home that I, in a manner, realised to the fullest extent the irreparable loss I had sustained and the inexpressible want I should now feel. I was alone, my wife and family being detained in Aberdeen for want of means to bring them home; and on entering the room—not thinking of the matter at the moment—a tremulous-like motion shot, like a flash of lightning, through my whole frame; and an audible expression, heard by the neighbours, escaped my lips as my eyes first encountered the bare walls. It appeared like a terrible dream, as I stood there, transfixed and immovable, with the door-sneck in my hand and my eyes fixed glaringly on those walls —walls which, but a few brief weeks before, had been adorned with the fruits of so many years' incessant toil, and unwearied assiduity, and which I held so dear, but now all gone! It was some time before I could persuade myself whether it was a delusion or a reality. The truth, however, forced itself upon me, and flinging myself upon a chair I gave full vent to my pent-up grief. By this means my feelings got relief.

"Everybody here seemed to know, after my return, what the consequences would be of my going to Aberdeen. Yes!—everybody knew how to make an egg stand on end when once Columbus had shown the way. Everybody, too, said that I had surely got a lesson now, and would cease to run about spending so much time after such useless and unprofitable things. Yes; a lesson I had, indeed, gathered—in every respect a dearly-bought one—but not the lesson they thought; and one which they were not very long in finding out.

"I did little or nothing that winter in the way

of collecting; but it only proved a lull before a tempest; for the fact, despite every opposition, developed itself in me during those few months that I could not live without these things, 'useless and unprofitable' though they were to others. And judging from what little I had already gleaned, that there was much more yet to be learned, I began to put my things in order.

"The spring had scarcely begun when I was at it again, and with redoubled energy and perseverance, and with ardour keener than ever, and quite regardless of what people said or might say. Conscious of my integrity, and the nobleness of the object I had solely in view, I was careless of the upas-tree-like venom of the slanderer, as well as the more subtle, though perhaps equally poisonous, tongue of the envious. I feared neither.

"I again pursued the same course as before, with this exception, that I did not go so often for whole nights. And, in addition to my former equipment, I now had a book for plants; a box, about nine inches square, for the more fragile of the insect tribe; something like a small trowel for digging up plants and chrysalides; and a hammer for chipping off any piece of rock I might wish to have and for splitting up fossils. I had, besides, something like a dredge-net, but to give it a name, or anything like a name, I cannot. This 'something' I used for the shallower parts of the sea shore and rock pools; I could either push or drag it, as circumstances required.

"By 1853 I had another good collection, and although not quite so numerous as my last, it contained more rare birds, insects, zoophytes, and plants—all preserved in a superior style.

"Dishealth now entering my family, I was under the melancholy and unavoidable necessity of disposing of this too. It was a heavy stroke, to say the least of it; but there was no help. It could not be avoided. From seventy to eighty cases were sold at this time, chiefly of birds, insects, and eggs; and nearly 300 specimens of nicely-preserved mosses and lichens, with about 200 zoophytes. Amongst the mosses and zoophytes were many which I could not get named, being new or undescribed.

"One individual bought all my first collection; but this one had many purchasers, and was scattered far and wide. Many of the cases went to England, not a few to Australia, and a number to America. The mosses and zoophytes are now in Australia; at any rate, the gentleman who bought them went there about six years ago.

"Having got the better of our affliction, I instinctively began again; for it was incorporated in me to the very heart's core, and could not be rooted out. I had now got a better gun—one which I had no occasion to hide below my coat-tails on meeting anybody or in passing through the town, and which I could depend more upon, being much superior to my old ricketty piece. I had now also a good shot-belt and powder-flask, all of which added considerably to my alacrity in procuring what I had to shoot.

"I now constructed traps, or something of that nature, of every imaginable form, size, and shape, and of every conceivable material which I thought would draw either the fancy or elicit the attention of life in whatever form it might appear. These were placed everywhere but in public roads; in water—fresh, salt, running, and stagnant; on land, in holes, in fields, in weeds, hung on trees and hidden in old dykes and stone-quarries. Some were visited daily, others once a week; whilst those in water, in mosses and marshy places, were only looked at once a month. By such helps and means I gained much information and obtained many, many things which I could never have procured in any other way. I likewise found much in the stomachs of fishes, which I looked more to now than formerly, and amongst the refuse of fishermen's lines, which I never would have seen or got otherways; and I never passed a dead animal without first making a very careful and diligent search of and about it. If it lay in an exposed situation I always took the liberty, if it would lift, of removing it to some more sheltered spot, and never failed to visit it as long as practicable afterwards.

"Thus by continual labour and incessant perseverance, I had again, in a few years, another and a most excellent and rich collection. But my own health now giving way, I was again compelled, with much reluctance, to put this—or at least a part of it—my third collection, also to the hammer. This occurred from 1858 to 1861–62. About sixty cases were parted with at this time, besides many things not in cases, of which a number went to the British Museum. Many of these cases, as well as those of my second collection, contained from one to three pairs of birds each—male and female—with their nests, containing either eggs or young, and were the admiration of all who saw them.

"On recovering again I found, to my unutterable sorrow, that, although my mind was as vigorous and as anxious and willing as ever, my ability otherwise was completely gone. I learned the sad and vexatious truth, which I had never before dreamed of, that a remarkably healthy, though not a robust, constitution, and one which seemed to defy everything, had at last given way, having been completely undermined by over-exertion and sapped to its very foundation by too much night exposure. I had the will and the desire still, but, alas! I had not the same power as before.

"Although obliged thus to relinquish my gun and its accompaniments, and to drop all night and early morning work, I still continued for a little to prosecute my marine researches. But this, too, had ultimately to be given up. I might have persisted longer, but being very seriously advised and warned to desist in time and not too foolishly, or rather too madly, dare the consequence which would in all probability soon have followed such a course of reckless and unwarrantable folly, I at last succumbed to these friendly admonitions, and relinquished this too. I had now eleven of a family—ten girls and a boy—to provide for; and this, of itself, was quite enough for any journeyman shoemaker, even although he had a thrifty wife. This was another and a most essential inducement to me to cease the one and pay more attention to the other work I had to do, so long as I was able.

"Notwithstanding the many thousands of objects from the animal and vegetable kingdoms, and the hundreds of cases, that I have from time to time been obliged to part with, I have still over two thousand of the former, including plants, and above sixty of the latter.

"The number of distinct species I have had, or met with, can never be ascertained; but it must have been immense.

"Though my field of labour may be considered by many to have been rather circumscribed, being chiefly confined to the county of Banff, yet it was quite enough—enough for any man, even with more time at his disposal than I had, and with more of the 'needful' at his command than ever I could boast of.

"Many of my discoveries have already—some long ago—become part of history; but a large proportion I am sure never will be known, or, if they are, they will in all likelihood never be assigned to their legitimate source, nor the true locality known where the objects were first found. I allude more particularly to some shrimps I met with, and to my insects, zoophytes, and plants. This may be thought to be rather premature speaking. It may. But if the ghosts of by-gone facts have anything to do with the shadows of coming events, and these things true, then I do not think I can be very far wrong. I will give two illustrations of what has already occurred, and leave the shadows to speak for themselves, when they come.

"Many years ago I used to send lots of crustacea to an acquaintance, to get named; but he, not knowing the subject himself, forwarded them to a gentleman who did. It happened that there was at least one new shrimp in one of these cargoes. The result was, a description of it soon appeared, for the edification of the natural history public, with the animal named, not after the *finder*, but in honour of the *sender;* thus completely ignoring both the locality and the discoverer.

"Again: some time since, a shrimp found on the Continent was named, and a description of it published, as a new species. This was all right, when the finder and publisher knew no better, and just what should have been done. But, the truth is, I had discovered the species here, in some abundance, about seven years before, and had transmitted specimens to a gentleman, and had received more than one letter from him saying that they were *new.*

Here, again, Banff and the old 'snob' had to resign their honours, now to a distant rival, even although the palm of victory was theirs by the seven years' priority of claim. Why the gentleman here did not publish the fact at the time I sent him the species—and he never did—is more than I can tell. However, we will say no more about it, but rather put it beside the shadows of coming events and let it loom there in the distance.

"Besides the great number of insects which I have been now and then forced to part with, I on one occasion lost 916 in one week that nobody was ever the better of. They were contained in twenty cases, each a foot square. Having filled these cases with the insects, newly arranged and numbered, I laid them aside in a garret until I could get glass. They were laid one above another crosswise, with the open side downwards, so as to keep out the dust. Six days afterwards I would begin to glaze them. On lifting the top case I was rather horrified at finding, not *insects,* but merely *pins,* with here and there a head or a wing. I lifted the others and found all the same. A more complete work of destruction and a better clean-out I never witnessed. If anyone had been hired to do the job he could not have done it better. Whether it had been mice or rats, I cannot tell; I never knew; but whichever or whatever it might have been, they did their business to a nicety unparalleled in the annals of natural history, or perhaps in any other annals.

"I have said that the specimens were all numbered. Each insect had its own number, and each box was numbered separately from its neighbour. I had this done in order that I might prepare a catalogue at some future date. I knew that there were sheets of numbers sold for the purpose, but, as I could not afford them, I got a lot of old almanacs, out of which I cut the figures. It was a long and tedious task, but patience and perseverance brought me through here too, as they had done in innumerable instances before.

"One other incident, of my child-life, before I finish. Though it happened when I was only about five years of age, from its being long after, and often, the theme of conversation at home, it is nearly, if not altogether, as fresh in my memory now as the day when it was enacted.

"I was, or rather we—for there were four of us, three other young chaps, like myself—were in a wood one day, seeking nests, and whatever else might turn up. We were separated a little, when one of the party called out, 'A byke! a byke! stickin' on a tree an' made o' paper.' This, of course, drew in all hands with a rush; for a byke was, at all times, a glorious capture, for the sake of the honey, besides the fun one had in skelping the bees, in order to get at it. But, before we reached the spot, a yell from the youngster of another sort, and one that made the whole wood resound again, of 'Oh! I'm stung; I'm stung. Oh! my lug. Oh! Oh!' arrested our progress. Seeing him take to his heels, blubbering loudly, we also ran; so that the retreat became general. When we had gone some distance, however, and there being no appearance of the enemy, a halt was resolved on to ascertain the true state of affairs, the strength of the foe, and so forth. All that could be learned, however, was simply that the byke was on a tree, 'made o' paper, and full o' yellow bees.' This so excited my curiosity—never having seen anything of the kind, that I at once proposed going back, to take it down, and home; but was met with a decided refusal by all. As I persisted in my determination to return, my three companions fled; so I was left alone in all my glory. Back, however, I went, nothing daunted.

"I soon found the byke, which was hanging to the underside of a branch of a tree, and very like a paper bag; and, on reaching to take it down, a bee settled on my hand and stung one of my fingers. I did not like this, for it was very sore, and, of course, drew back in case of more, sucking at and blowing upon my finger all the time.

"Having once seen the thing, to leave it was impossible. That was no trait of mine. Help or no help, I could not go without it. What was to be done? True, I had a something on my head which might at one time, with all propriety, I dare say have been called a *bonnet*, but now it was rather of a kin to the good wife's riddle—it was '*a' holes.*' Besides, what did remain was too hard; it would have broken the byke, and that I wished home whole; so it was of no use. I was barefooted; so that a stocking was out of the question. What then?

"How long I had puzzled my little brains I cannot say; it must have been long—no doubt I sucked my finger all the while—as it was getting dark before I hit upon a plan. It was a very easy matter and would cost but little trouble, then, to take off my shirt; and this, although not without its ventilation, like my head-piece, still was not quite so bad as the good wife's riddle.

"With this I again approached, now cautiously, and succeeded in getting it all round; and I then managed, somehow, to detach the byke from the branch. This done, I tucked up all the corners, fastened them into a knot, so as to keep all in that was in, and then flew as hard as I could in case of interference from outsiders, and reached home in safety.

"I peeped in at the keyhole to see if the coast there was clear; but—no! my father was in. I had an old iron pail in a recess on the stair in which I kept certain things; here my shirt and its contents were safely deposited to await next morning.

"In I went now, as if nothing had happened, took my supper, and in a little while was ordered to bed with the rest. Well, I had just taken off my jacket, without even thinking what I was about, when one of my young friends cried out,—

" 'Mither, mither, Tam hasna' on his sark.'

" 'Where's your shirt, sir?' said my father.

" 'Dinna ken.'

" 'What? Give me down my razor-strap.'

" I knew one virtue that strap had, for it and I were old acquaintances.

" 'Now, sir, tell instantly where it is, or what you have done with it.'

" 'It's in the *bole*, on the stair.'

" 'Go and bring it here immediately.'

" There was no help. I had to go!

" 'What have you in it?' my father asked on my return.

" 'A bees' byke.'

" 'A what?' shouted father and mother, in one breath.

" 'A bees' byke,' I repeated.

" 'Did I not tell you,' said my father, 'only the other day, not to bring any more of those things here, endangering people with them? Besides, to think of you stripping yourself naked in a wood for such a purpose!'

" 'But this ane's made o' paper.'

" 'Made o' fiddlesticks!'

" 'Na; I'll let you see't.'

" 'I dare say you will. But I don't want to see't. Let it alone, this instant, I tell you, and go away to bed; or I'll show you something that will do you more good and be less dangerous.'

"On asking my mother next morning for my byke she told me that my father had caused her to soak my shirt in boiling water for about an hour, and, then opening it, she had scraped all off—bees and all—and had thrown all out. She also told me that, instead of being a byke, it was *a wasps' nest.*

"A few words more, which may be of use, and, if not, no matter.

"Although come of Scotch parents I am, myself, English by birth, having been born in Gosport. This took place on the 25th December, 1813; but *the precise hour I do not remember.* The reason I came into the world there is simply this, I suppose: my father was among the Fifeshire militia, and they were there at that time; and my mother being allowed to accompany my father she was there with him. The regiment was disembodied some time after; and my mother being an Aberdonian, my father—a Fife man himself—came to Aberdeen with her, where he lived and died.

"I, myself, lived there until I was nineteen, when I left and came to Banff; and here I am still.

"THOMAS EDWARD.

"*Banff, May,* 1873."

MATHEMATICIANS IN HUMBLE LIFE.

A BRIEF notice of the mathematicians who flourished in the North of England, and particularly in the manufacturing districts of Lancashire, about the close of the last and the early part of the present century, cannot fairly be omitted in treating of scientific men in humble life. The Mathematicians of Lancashire were as remarkable a class of men, in their way, as the Naturalists. "The pages of our mathematical periodicals," said Mr T. T. Wilkinson, in a paper read before the Manchester Literary and Philosophical Society, 1854, "will readily furnish more exquisite specimens of the geometry of the Greeks, mostly produced by Lancashire operatives, than could be found in many laboured treatises of those whose official situations have led them to cultivate the subject."

Further testimony as to the character of these humble mathematicians was borne by Mr. G. Harvey, F.R.S., in a communication which he made to the British Association, at its first meeting, in York. In this, he says: "It was my desire to have drawn the attention of the meeting to the very remarkable circumstance of the geometrical analysis of the ancients having been cultivated with eminent success in the northern counties of

England, and particularly in Lancashire. The proofs of this may be gathered from the variety of periodical works devoted almost exclusively to this lofty and abstract pursuit. I have now before me several beautiful specimens of the geometry of the Greeks, produced by men in what, for distinction's sake, we call the inferior condition in life. The phenomenon — for such it truly is — has long appeared to me a remarkable one, and deserving of an attentive consideration. Dr. Playfair, in one of his admirable papers in the 'Edinburgh Review,' expressed a fear that the increasing taste for analytical science would at length drive the ancient geometry from its favoured retreat in the British Isles; but at the time he made this desponding remark the professor seemed not to be aware that there then existed a devoted band of men in the North, resolutely bound to the pure and ancient forms of geometry, who, in the midst of the tumult of steam engines, cultivated it with an unyielding ardour, preserving the sacred fire under circumstances which seem, from their nature, most calculated to extinguish it. In many modern publications, and occasionally in the senate-house problems proposed to the candidates for honours at Cambridge, questions are to be met with derived from this humble and honourable source. The true cause of this remarkable phenomenon I have not been able clearly to trace. The taste for pure geometry— something like that for entomology among the weavers of Spitalfields—may have been transmitted from father to son, but who was the distinguished individual first to create it in the peculiar race of men here adverted to seems not to be known. Surrounded by machinery, with the rich elements of mathematics in their most attractive forms, we should have imagined that a taste for mathematical combinations would have exclusively prevailed, and that inquiries locked up in the deep and, to them, unapproachable recesses of Plato, Pappus, Appolonius, and Euclid would have met with but few cultivators. On the contrary, porisms and loci, sections of ratio and of space, inclinations and tangencies—subjects confined, among the ancients, to the very greatest minds—were here familiar to men whose condition in life was, to say the least, most unpropitious for the successful prosecution of such elevated and profound pursuits."

A society of mathematicians was established in Manchester as early as the year 1718, which embraced several distinguished names. Their influence and example resulted in the formation of a mathematical society at Oldham, in 1794, by Messrs. Abraham Jackson, William Hilton, William Travis, James Travis, and John Bardsley, whose members soon distanced those of the parent society in their attainments.

"The establishment of this society"—we quote from Mr. Wilkinson's communication—"supplied the requisite impulse for the full development of this local geometrical taste; and no reasonable doubt can exist that the Manchester and Oldham mathematical societies were really the great promoters of the revival of the study of the ancient geometry in Lancashire, for during the latter half of the last century, and almost up to the present date, they have numbered amongst their members several of the most distinguished cultivators of ancient geometry in modern times."

The mathematical periodicals—'The Gentleman's Diary,' 'Burrows's Diary,' and some others—teemed with the productions of the Lancashire mathematicians, of whom Wolfenden and Butterworth were perhaps the most distinguished, the former being confessedly one of the most skilful cultivators of plane geometry in the last century. "Besides those who resided in the immediate neighbourhood of Oldham we find the names of Mabbott, Wood, Holt (*Mancuniensis*), Dr. Clarke (*Salfordiensis*)—subsequently of the Royal Military College, Sandhurst, and author of the 'Perspective,' 'Rationale of Numbers,' &c.—as then resident in Manchester, and in constant communication with Crakelt, Hutton, Lawson, Wildbore, and other editors of mathematical periodicals; nor can it be doubted from the evidence of existing documents, that the predilection evinced for the study of the ancient geometry by these active members of the Lancashire school exercised considerable influence upon the pursuits of such able and proficient mathematicians in other localities as Campbell, Cunliffe, Dalton, Davies, Gough, Lowry, Ryley, Whitby, Shepherd, and Swale."

The predilection for geometrical pursuits among the operatives of the North is accounted for by the nature of their employment. On this point Mr. Wilkinson observes: "The weaver at the loom, the farmer in the field, the mechanic in the shop, or the miner in the drift is too often occupied by manual labour to be able to write out long analytical investigations, but each can contemplate and deduce at pleasure the properties of a geometrical diagram either actually constructed or mentally conceived. The farmer and the miner soon acquire the power of depicting vivid mental representations of the constructions necessary for their geometrical inquiries, and are thus enabled to carry on their processes of deduction even when buried in the mine or following the plough. The weaver and the mechanic can sketch their diagrams on the slate, and thus pursue their favourite studies whilst their hands and feet are almost instinctively engaged in their monotonous occupations. In this manner many of the speculations of the geometers of the Lancashire school are known to have been conducted. Nor is the practice at present by any means extinct; most of the profound investigations of Wolfenden and Butterworth which appeared in the 'Mathematical Companion' and elsewhere were composed, almost word for word, while contemplating the requisite diagrams suspended from the framework of their looms." How closely did the practice of these men correspond with that of the operative botanists, many of whom mastered the intricacies of the Linnæan classification by a similar process and under similar disadvantages!

Wolfenden and Butterworth, resided throughout their long lives in the neighbourhood of Oldham, and both belonged to the society of mathematicians established in that town. Both, we are assured, were nursed and reared in poverty; both had to struggle through life with the difficulties which such a position involved; and both, in old age, would have had to endure its keenest privations but for pecuniary assistance rendered to them by members of the Manchester Literary and Philosophical Society.

When the Society for the Relief and Encouragement of Scientific Men in Humble Life was established at Manchester in 1843, Butterworth (who was locally known as "Jack o' Bens") lived at Haggate, near Oldham. Mr. Binney, one of the most active promoters of that society, made a pilgrimage to Haggate in search of him. How he found the old man is thus told:—He made many inquiries in the neighbourhood for John Butterworth, and was directed to one of that name who was a shopkeeper, another who was a beerseller, and a third a schoolmaster—but not the right one. On stating that the person he wanted was a mathematician, a woman said, 'Oh, it's owd Jack o' Bens,' and immediately showed him the house. He found Butterworth in a small narrow room, about two yards in width, where he taught a few children, by means of which occupation chiefly he supported himself.

Butterworth was a fine old man, having an extraordinarily massive head covered with snow-white hair, and a countenance beaming with intelligence and good nature. His manners were simple and very pleasing.

The early history and later achievements of Butterworth are thus told by Mr. Binney:—"He was born at Haggate, in 1774. When three or four years old he went to a school kept by an old woman, and afterwards to a free school, but he was taken away before he learned to read; and at six years of age was sent to work at a Dutch wheel in Royton, where he obtained 1*s.* 4*d.* per week wages. He was afterwards employed in winding bobbins, and at ten years of age began to weave. He was between fifteen and twenty years old before he could read and write well.

"Butterworth and three other young men were accustomed to meet for mutual instruction. His first taste for mathematics was obtained from reading an old almanac. He then associated with some who studied geometry and took delight in it. About 1790 there was a Mathematical and Philosophical Club at Oldham, and Butterworth, in company with the late Mr. Wolfenden and others, became a member of it. From this society he derived many valuable books, which he read with great avidity, and soon began to answer mathematical questions in public prints. From the commencement of the nineteenth century to the present time (1843) he was a regular contributor to nearly all the mathematical publications of the day. The 'Mathematical Companion'—many prizes of which he obtained—the 'Lady's Diary,' and the 'Gentleman's Diary,' and other works bore testimony to the value and extent of his contributions. Many years ago he read and thoroughly mastered Newton's 'Principia,' and, by the most competent judges he was considered to have a profound acquaintance with many of the works of the old geometricians. After the death of his father he continued, until 1837, to reside with his mother." She died in that year.

Butterworth's wages, as a weaver of fustian, seldom exceeded 15*s.* a week; but in his later years he did not earn more than 8*s.*, and he found it necessary to eke out a living for himself and to support his aged parent by keeping a school. He also sold sweetmeats. His school, at the date of Mr. Binney's visit, before referred to, consisted of four small children, as day scholars, for each of whom he received 2½*d.* per week; and six factory children, who came to him for instruction in the evenings, and who paid him 3*d.* per week.

Notwithstanding this miserable pittance, and being much affected with a disease of the lungs, Butterworth was far from complaining; indeed he was comparatively contented with his humble lot. He mentioned, as a source of annoyance to him, being obliged by his poverty to answer mathematical questions brought to him by persons who wished to insert them in newspapers as their own, and who paid him sixpence or a shilling for his services. Not that he cared for the credit of the solutions, but, he said, having spent a lifetime in the investigation of truth, it was hardly right even to allow people to deceive the public by claiming for themselves a knowledge they did not possess.

In private life, whether we consider his duties to his parents or his neighbours, Butterworth was a most worthy and upright man. 'Leybourn's Mathematical Repository,' to which Mr. Lowry, of the Royal Military College, and many of the most eminent mathematicians of that day, contributed, contained abundant evidence of the sort of questions that he was called upon to elucidate. The following is a specimen:— "Given the base of a spherical triangle upon the surface of a sphere, and the area of the triangle— to find the locus of the vertex?" This problem, amongst many others, Butterworth solved. The solution printed in the volume was by Mr. Lowry; but Butterworth's name was attached as one of the solvers. "He was confessedly," says Mr. Wilkinson, "the prince of Lancashire geometers, both for variety and extent. Were the whole of his correspondence to be collected it would form several bulky volumes on geometrical analysis. The construction of triangles from given data would, indeed, form a conspicuous feature of such a collection; but properties of the conic sections, and triangles, problems in tangencies and loci, sections of ratio and of space, inclinations and porisms would appear at intervals in considerable variety, and mostly treated in a style which, for originality and elegance, would certainly have done no discredit to either the author of the 'Elements' or 'the great geometer of Perga.'"

From 1843—when he was nearly seventy years of age—until the close of his life, Butterworth received an allowance from the relief society before mentioned, for which, small though it was, he was extremely grateful. As a testimony of their respect for the worthy old man, Sir W. Fairbairn, Bart. (then Dr. Fairbairn), Mr. E. W. Binney, and other gentlemen from Manchester, attended his funeral.

It is a matter of regret that the fame of the mathematicians has not, like that of the naturalists, in humble life, been handed down, in some measure, to their successors of the present generation.

LONDON: PRINTED BY WILLIAM CLOWES AND SONS, STAMFORD STREET AND CHARING CROSS.

www.ingramcontent.com/pod-product-compliance
Ingram Content Group UK Ltd.
Pitfield, Milton Keynes, MK11 3LW, UK
UKHW010337140625
459647UK00010B/665

* 9 7 8 1 1 0 8 0 3 7 9 0 7 *